本教程获得

中国学位与研究生教育学会立项
(B1-2013Y03-005) 基金资助
北京理工大学 2015 年度研究生
教学团队立项建设基金资助

北京理工大学"双一流"建设精品出版工程

科技英语交流
（第2版）

Scientific Communication in English
(2nd Edition)

主编　王玉雯　李　恒　柳君丽
编者　赵　蓉　马文艳　郑　群　高　波
　　　许子艳　张俊梅　吕　行　张　帆
　　　王　锦　张　毓

北京理工大学出版社
BEIJING INSTITUTE OF TECHNOLOGY PRESS

内容简介

本教程以国际顶级期刊科技英语论文为素材，详细分析了科技英语论文的写作构成要素和语言特点以及撰写技巧，设计了相关的练习，旨在帮助学习者掌握科技英语论文写作特点和技巧，提高科技英语交流能力。

版权专有　侵权必究

图书在版编目（CIP）数据

科技英语交流 / 王玉雯，李恒，柳君丽主编 . —2 版 . —北京：北京理工大学出版社，2018.12（2020.8重印）

ISBN 978-7-5682-6526-3

Ⅰ . ①科… Ⅱ . ①王… ②李… ③柳… Ⅲ . ①科学技术-英语-论文-写作 Ⅳ . ①G301

中国版本图书馆 CIP 数据核字（2018）第 282862 号

出版发行 /	北京理工大学出版社有限责任公司
社　　址 /	北京市海淀区中关村南大街 5 号
邮　　编 /	100081
电　　话 /	（010）68914775（总编室）
	（010）82562903（教材售后服务热线）
	（010）68948351（其他图书服务热线）
网　　址 /	http：//www.bitpress.com.cn
经　　销 /	全国各地新华书店
印　　刷 /	北京虎彩文化传播有限公司
开　　本 /	787 毫米×1092 毫米　1/16
印　　张 /	9.75
字　　数 /	235 千字
版　　次 /	2018 年 12 月第 2 版　2020 年 8 月第 2 次印刷
定　　价 /	39.00 元

责任编辑 / 梁铜华
文案编辑 / 梁铜华
责任校对 / 杜　枝
责任印制 / 王美丽

图书出现印装质量问题，请拨打售后服务热线，本社负责调换

前言（第 2 版）
PREFACE (Second Edition)

《科技英语交流》2015 年在北京理工大学出版社出版发行后，受到广大读者的欢迎和推崇，并被评为北京理工大学校级"十三五"（2018 年）规划教材。

为创造混合式教学途径，本教材编委在 2018 年《科技英语交流（第 2 版）》增加了如下内容：

首先，为帮助学习者提高科技英语论文写作效果，我们依据二语习得理论设计了《科技英语交流》写作微课，精心提炼教材的核心内容，撰写了九个单元的微课课件，请美籍教师 Brooke Holliday 为微课课件录音，陈天予编辑了部分录音文件。该微课课件丰富了科技英语论文写作的教学模式，学习者既能在阅读中学习科技英语论文写作技巧，而且能运用听觉加强科技英语论文写作技能。录音文件及文字可通过扫描教材里的二微码得到。

李恒组织编者编写了九个单元的微课课件，其中，北京理工大学王玉雯编写了第一章和第四章，李恒编写了第五章和第十二章，赵蓉编写了第六章和第七章，马文艳编写了第二章，柳君丽编写了第三章，高波编写了第八章。李恒审核了课件录音。

其次，王玉雯阅读全册，修改了个别的语言错误，并组织编者编写了教学课件 PPT（如果需要，可在北京理工大学出版社网站免费下载，网址：www.bitpress.com.cn）。其中，北京理工大学王玉雯编写了第一章、第四章和第十一章，李恒编写了第五章和第十二章，赵蓉编写了第六章和第七章，马文艳编写了第二章，柳君丽编写了第三章，高波编写了第八章，许子艳编写了第十章，吕行编写了第十三章，中国科学院大学郑群教授编写了第九章。

本教材得到了北京理工大学研究生院 2017 年多模态科技英语交流的研究课题（李恒负责）和 2018 年北京理工大学研究生明星课程建设（王玉雯负责）的资助，在此，我们表示衷心的感谢！

基于《科技英语交流》教材设计的慕课于 2019 年 12 月在学堂在线上线，"科技英语交流"慕课配有教学视频、课程 PPT、单元练习和答案，可免费学习，网址为：https://next.xuetangx.com/course/bitP05021001988/1515925。

王玉雯

2020 年 7 月 12 日

前言

科技英语交流能力决定了科研人员能否成功地在国际期刊发表他们的研究成果，能否有效地与同行进行交流。而国内外许多研究表明，中国学习者的主要问题是不了解英语科技论文的语言特征，缺乏语篇衔接知识和技能，难以驾驭科技英语词汇的运用。

为此，我们经过六年的教学和科研探索，编写了《科技英语交流》教程，旨在帮助研究生和科研人员提高科技英语交流能力。该教程分为两个部分，第一部分为科技英语论文写作和投稿，共十章。第一章至第八章分别描述了科技英语论文写作特点，论文标题、摘要、引言、方法、结果、讨论和结论的写作，从宏观和微观结构层面陈述科技英语论文写作各个部分的构成要素和写作技巧，详细分析科技英语顶级期刊论文的写作结构要素和语言特征，以帮助学习者熟悉并掌握科技英语论文写作的基本要素和技巧。第九章为论文修改方法，第十章为论文投稿技巧。第二部分（第十一章至第十三章）为国际会议英语交流。

在撰写该教程期间，我们得到了澳大利亚阿德莱德大学玛格丽特·卡吉尔教授的帮助和指导。她帮我们审阅了初稿，提出了详细的修改建议。中国科学院苏州纳米技术与纳米仿生研究所王锦副研究员提供了大量的参考素材，并提出了编写建议。在此一并致谢。

《科技英语交流》教程作为"学术英语交流能力培养模式的研究"课题成果之一，该课题由中国学位与研究生教育学会批准立项（B1-2013Y03-005），2015年8月通过专家评审，已结题。课题组成员2013年开始着手编写《科技英语交流》教程，2014年10月完成初稿，经多次反复修改才得以问世。该课程2010年被评为北京理工大学研究生重点骨干课程建设，2015年被列为北京理工大学研究生教学团队建设内容之一。

本教程作者都是一线英语教师，其中北京理工大学王玉雯撰写第一章和第四章，马文艳第二章，柳君丽第三章，赵蓉第六章和第七章，高波第八章，许子艳第十章，李恒第十二章，吕行第十三章；中国科学院大学郑群撰写第九章；华北电力大学张帆编写第五章；北京工业大学张俊梅第十一章；中国科学院苏州纳米技术与纳米仿生研究所王锦撰写了科技英语翻译的启示；北京航空航天大学博士研究生张毓编写了第一章的衔接手段。

  本教程适合有志于在国际期刊发表科技论文的科研人员和在校研究生。科技英语交流在北京理工大学的教学学时为54，需要教学课件的老师可联系wyw@bit.edu.cn。

目 录
CONTENTS

Chapter 1　Features of Scientific Writing ……………………………………… (001)
- 1.1　Clarity ……………………………………………………………………… (001)
- 1.2　Contribution ……………………………………………………………… (006)
- 1.3　Documentation …………………………………………………………… (006)
- 1.4　Language Use …………………………………………………………… (008)
- 1.5　Tips on Effective Scientific Writing …………………………………… (009)
- 1.6　Summary ………………………………………………………………… (012)

Chapter 2　How to Write a Title ……………………………………………… (014)
- 2.1　Writing Principles: BAD Principles …………………………………… (014)
- 2.2　Common Features ……………………………………………………… (016)
- 2.3　Specific Analyses ……………………………………………………… (018)
- 2.4　Summary ………………………………………………………………… (020)

Chapter 3　How to Write Abstracts ………………………………………… (021)
- 3.1　Basic Components ……………………………………………………… (021)
- 3.2　Specific Analyses ……………………………………………………… (022)
- 3.3　Tips on Writing a Good Abstract ……………………………………… (031)
- 3.4　How to Build Your Own Syntactic Templates ……………………… (034)
- 3.5　Summary ………………………………………………………………… (035)

Chapter 4　How to Write Introduction ……………………………………… (037)
- 4.1　Basic Components ……………………………………………………… (037)
- 4.2　Specific Analyses ……………………………………………………… (038)
- 4.3　Tips on Writing a Good Introduction ………………………………… (048)
- 4.4　Common Syntactic Templates ………………………………………… (049)

4.5　Summary ……………………………………………………………（050）

Chapter 5　How to Write Methods …………………………………………（052）
5.1　Basic Components …………………………………………………（052）
5.2　Specific Analyses ……………………………………………………（053）
5.3　Tips on Writing Methods Section …………………………………（061）
5.4　Use Proper Syntactic Templates as Well as Useful Expressions ……（062）
5.5　Summary ……………………………………………………………（063）

Chapter 6　How to Write Results …………………………………………（064）
6.1　Basic Components …………………………………………………（064）
6.2　Specific Analyses ……………………………………………………（065）
6.3　Tips on Writing a Good Results ……………………………………（069）
6.4　Syntactic Templates in Writing the Results Section ………………（072）
6.5　Summary ……………………………………………………………（073）

Chapter 7　How to Write Discussion ………………………………………（074）
7.1　Basic Components …………………………………………………（074）
7.2　Specific Analyses ……………………………………………………（075）
7.3　Tips on Writing a Good Discussion ………………………………（080）
7.4　Syntactic Templates in Writing the Discussion Section ……………（081）
7.5　Summary ……………………………………………………………（082）

Chapter 8　How to Write Conclusion ……………………………………（083）
8.1　Functions and Components of Conclusion ………………………（083）
8.2　Specific Analyses ……………………………………………………（084）
8.3　Strategies for Creating Effective Conclusions ……………………（087）
8.4　A Sample Conclusion for Revision (Table 8.9 and Table 8.10) ……（090）
8.5　Summary ……………………………………………………………（091）

Chapter 9　How to Revise a Scientific Manuscript ………………………（093）
9.1　The Definition of Revision …………………………………………（093）
9.2　The Peer Review and Revision ……………………………………（093）
9.3　Three *Myths*（误区）about Revision ……………………………（096）
9.4　Tips to Revise the Draft/Research Paper …………………………（097）
9.5　Resubmission after Revision ………………………………………（102）

Chapter 10　How to Submit a Manuscript ………………………………（104）
10.1　How to Prepare a Manuscript ……………………………………（104）

10.2　Where to Submit ……………………………………………………… (104)
10.3　What to Submit ……………………………………………………… (105)
10.4　How to Submit ……………………………………………………… (108)
10.5　Understanding Review Process ……………………………………… (108)
10.6　Dealing with Rejection ……………………………………………… (110)
10.7　Summary ……………………………………………………………… (110)

Chapter 11　International Conference ……………………………………… (112)
11.1　Kinds of Meetings …………………………………………………… (112)
11.2　Conference Documents ……………………………………………… (112)
11.3　Conference Activities ………………………………………………… (117)
11.4　Conference Stage …………………………………………………… (117)
11.5　Ways to Present Papers ……………………………………………… (118)
11.6　Conference Abstract ………………………………………………… (118)

Chapter 12　How to Deliver Presentations ………………………………… (120)
12.1　Features of an Effective Presentation ……………………………… (120)
12.2　Tips on an Effective Presentation …………………………………… (120)
12.3　The Question-and-answer Session …………………………………… (125)
12.4　Summary ……………………………………………………………… (127)

Chapter 13　How to Create Scientific Posters ……………………………… (128)
13.1　Basic Components …………………………………………………… (128)
13.2　Format ………………………………………………………………… (134)
13.3　Software ……………………………………………………………… (137)
13.4　Presenting …………………………………………………………… (137)
13.5　Summary ……………………………………………………………… (138)

References ……………………………………………………………………… (139)

附录　翻译的启示 …………………………………………………………… (143)

Chapter 1
Features of Scientific Writing

Scientific writing has different definitions. For example, Day and Gastel (2007:3) have defined scientific writing as to denote the reporting of original research in journals, through scientific papers in standard format (IMRaD, a short form for Introduction, Methods, Results and Discussion).

We begin by providing an overview of some important features of scientific writing such as clarity, contribution, documentation and language use, and then we offer tips for effective writing.

1.1 Clarity

The first feature of scientific writing is *clarity*（清晰）involved in its macro organization and micro organization. The macro organization refers to external organization such as chapters, sections, and paragraphs in the *dissertation*（学位论文）and IMRaD or AIMRaD (a short form for Abstracts, Introduction, Methods, Results and Discussion) in a journal paper whereas the micro organization indicates that scientific writing might organize information by employing general-specific and problem-solution patterns. The micro organization also involves the connection between paragraphs and between sentences.

1) IMRaD or AIMRaD

The IMRaD or AIMRaD can be a macro organization in a research paper. The logic of IMRaD can be defined in the following questions (Day and Gastel, 2007:10):

(1) What question (problem) was studied?
(2) How was the problem studied?
(3) What were the findings?
(4) What do these findings mean?

AIMRaD refers to a *format*（格式）with abstracts, introduction, methods, results and discussion in a journal paper. However, as a scientific writer, you should recognize that, "within a common core structure, there are variations from field to field and from journal to journal: always check the specific requirements of your *target journal*（投稿期刊）" (Cargill and O'Connor, 2013: 11) before writing the research work.

2) The general-specific pattern

The *general-specific*（总述—分述）pattern is quite popular and is a micro organization in scientific writing. The general-specific texts move from broad statements to narrower ones. However, they often widen out again in the final sentence (Swales and Feak, 2012:33). Swales and Feak (ibid 同上) also suggest that this pattern text usually begins with one of the following elements:

(1) A short or extended definition,

(2) A contrastive or comparative definition, or

(3) A generalization or purpose statement.

The following paragraph is an example of the general-specific text taken from the journal *Macromolecules* (高分子) written by Carter et al. (2005:4595). This paragraph begins with a generalization statement and then more specific information is provided.

E. g. 1: *Branched polymers owe much of their utility to the presence of a large number of chain ends per molecule and their chain architecture. The latter can have a profound effect on materials properties, such as rheology and solubility. On the other hand, the large number of chain ends can be used to add useful chemical functionality, which may differ from similar functionality added along the main chain. These polymers can be produced in chain growth polymerization by using branching monomers, which act as both monomers and transfer agents or as monomers and initiators. Both of these functions are combined in approaches that use addition-fragmentation as the branch forming reaction, and of these, the reversible addition-fragmentation chain transfer (RAFT) methodology introduced by Thang et al. also offers the opportunity to modify the end groups. This method has proved to be a useful technique for the synthesis of reactive polymers and polymers with well-defined architectures...*

The above paragraph is also a good case of *cohesion* (衔接), in which sentences are closely related by employing "the latter," "these," and "this" or by repeating the key words "chain ends."

Another pattern is the problem-solution pattern, which will be introduced in the next subsection.

3) The problem-solution pattern

The problem-solution pattern was first proposed by Hoey in 1983. This pattern usually has four parts as follows:

(1) Description of a situation,

(2) Identification of a problem,

(3) Description of a solution,

(4) Evaluation of the solution.

To better understand this pattern, we need to take a look at E. g. 2 below and identify the situation, the problem, the solution and the evaluation of the solution.

E. g. 2: *This paper considers pairwise formations between unmanned aerial vehicles (UAVs) where an agent gains a fuel benefit by flying in the wake of another (i. e., a reduction in aerodynamic drag). The objective of each UAV is to travel from source to target locations while consuming the least amount of fuel. UAVs are able to reach their destination alone, however, joining in formation can potentially improve their fuel economy. When agents are noncooperative, the potential benefits of flying in formation bring up the issue of fairly distributing the leader task. The goal of this paper is to find* *optimal cooperation-inducing leader allocations that minimize UAV fuel consumption and provide individual agents with algorithms to compute them. Such benefit-driven cooperation mechanisms are a necessary building block to realize the potential benefits of collaboration in groups of noncooperative agents bargaining over the possibility of teaming up.* (problem being investigated) *The results of this paper can be applied to scenarios involving bargaining and auctions, task allocation in teams, and transferable utility games.* (Richert and Cortés, 2013:3189)

E. g. 2 is a good problem-solution pattern, quite popular in introducing the background, research problem, solution to the problem and significance of the solution. To be more specific, the first three sentences described the research situation while the fourth sentence introduced the problem. The fifth sentence presented a solution to the problem. The last sentence revealed the application of the solution, which was the evaluation of the solution.

4) Cohesion

Cohesion is the main medium to organize the sentences into text, which shows the relationship between parts in a text beyond the control of a sentence structure (Zhang, 2000). "Cohesion occurs where the interpretation of some element in the discourse is dependent on that of another" (Halliday and Hasan, 2001: 4). Cohesion can be constructed through cohesive devices. As for the taxonomy of cohesive devices, Halliday and Hasan (2001) proposed grammatical cohesion and lexical cohesion. The grammatical cohesion can be further categorized as *reference* (照应), *substitution* (替代), *ellipsis* (省略) and *conjunction* (连接). It should be noted that substitution and ellipsis are seldom used in academic writing. Lexical cohesion consists of lexical *reiteration* (复现) and *collocation* (搭配). Table 1.1 and Table 1.2 provide the categories of cohesive devices and examples, which are frequently used in writing.

Table 1.1 Categories of grammatical cohesion

Grammatical Cohesion		Examples
Reference	*Personal reference* 人称照应	I, you, we, he, she, it, they, etc.
	Demonstrative reference 指示照应	this, these, that, those, here, there, the
	Comparative reference 比较照应	same, similar, other, different, else, better, etc.
Substitution	Nominal	one, ones
	Verbal	do
	Clausal	so, not

continued

Grammatical Cohesion		Examples
Ellipsis	Nominal	
	Verbal	
	Clausal	
Conjunction	*Additive* 递进	and, and also, moreover, furthermore, besides, etc.
	Adversative 转折	however, nevertheless, despite, but, yet, though, although, etc.
	Causal 因果	because, so, therefore, then, hence, for this reason, consequently, as a result, etc.
	Temporal 时空	first, secondly, finally, at last, in short, in a summary, in a word, briefly

Table 1.2 Categories of lexical cohesion

• Lexical cohesion	
• Reiteration	• Repetition (same word)
	• Synonym
	• *Superordinate* 上义词
	• General word
• Collocation	

To better understand the uses of cohesive devices in scientific paper writing, we presented findings in Zhang's research in chemical journal abstracts written by native English authors and Chinese authors, which was conducted in 2013. The results showed that the reference devices accounted for a larger percentage than the conjunction devices of the overall use of grammatical cohesive devices in both *corpora* (语料库). The overall occurrence of grammatical cohesive devices adopted in Chinese authors' abstracts was lower than that in native authors'. A significant difference existed in the use of grammatical cohesive devices between two groups. In terms of the distribution of personal reference, Chinese authors tended to distinctly *underuse* (使用不足) the first-person plural pronoun. The definite article "the" had the largest number in both corpora, but the occurrences of selective *nominal* (名词性) demonstratives "this" and "these" were much lower in the Chinese authors' abstracts than those in native authors'. As for the comparative reference, Chinese authors tended to use less adjectives of comparison and adverbs of comparison than native authors. In terms of the use of conjunctions, additive devices occurred far more frequently than other types of conjunctions in both corpora and Chinese authors were found to underuse adversative devices. Among the top ten conjunctions in each corpus, native authors tended to use more additive and adversative conjunctions while the most frequently used conjunctions by Chinese authors confined to additive and temporal items. Compared to native authors, Chinese authors tended to put more conjunctions in the initial sentence position, which needed to be improved in their future English writing. Thus, we hope the above findings can provide you with stimulating thoughts in your use of cohesive devices in your scientific writing.

Cohesion plays an important role in promoting the flow of scientific writing. *Flow* (连贯) is moving from one statement in a text to the next by establishing a clear connection of ideas, which is important to help your reader follow the text. Linking words and phrases can help a writer maintain flow and establish clear relationships between ideas. The following two examples present us the differences of the flow in scientific writing, in which E. g. 3 below is a failure and E. g. 4 below is a good illustration of the flow.

E. g. 3: *Lasers have found widespread application in medicine. Lasers play an important role in the treatment of eye disease and the prevention of blindness. The eye is ideally suited for laser surgery. Most of the eye tissue is transparent. The frequency and focus of the laser beam can be adjusted according to the absorption of the tissue. The beam "cuts" inside the eye with minimal damage to the surrounding tissue — even the tissue between the laser and the incision. Lasers are effective in treating some causes of blindness. Other treatments are not. The interaction between laser light and eye tissue is not fully understood.* (Swales and Feak, 2010:21)

E. g. 4: *Lasers have found widespread application in medicine. For example, they play an important role in the treatment of eye disease and the prevention of blindness. The eye is ideally suited for laser surgery because most of the eye tissue is transparent. Because of this transparency, the frequency and focus of the laser beam can be adjusted according to the absorption of the tissue so that the beam "cuts" inside the eye with minimal damage to the surrounding tissue—even the tissue between the laser and the incision. Lasers are also more effective than other methods in treating some causes of blindness. However, the interaction between laser light and eye tissue is not fully understood.* (Swales and Feak, 2010:21)

Both E. g. 3 and E. g. 4 have expressed the same idea, but they differ in terms of the flow. E. g. 3 failed in showing the relationship between sentences though sentences are grammatically correct whereas sentences in E. g. 4 are both grammatically correct and coherent. The flow has been achieved by the underlined words and phrases in E. g. 4.

In E. g. 4, "for example" is used to help readers distinguish the major idea (the first sentence "Lasers have found widespread application in medicine") from the supporting evidence whereas the pronoun "they" has substituted "lasers" to avoid repetition. The linking word "because" has successfully combined the third and the fourth sentences. The phrase "Because of this transparency," has been added to closely connect the preceding sentence with the following one. Another phrase "so that" has been employed to combine the two sentences into one. The underlined word "also" has contributed to the additive. Another underlined word "than" has effectively combined two sentences into one while the last underlined word "however" has been employed to show the change of ideas (adversative). Therefore, seven linking devices in E. g. 4 have effectively connected ideas between sentences and clearly showed readers the relationship.

E. g. 5: *Further, extra fixed effects terms can be incorporated to represent additional information or characteristics of the input image pair. These fixed effects are represented by the terms...*

In E. g. 5, the demonstrative reference "These" and the repeating key phrase "fixed effects" have effectively joined the idea of the two sentences together. Thus another way to maintain flow is to repeat the key words of the two sentences. E. g. 6 is *a good case in point*（一个恰当的例子）.

E. g. 6: <u>*Model I*</u> *is a first step towards building a framework that incorporates image quality effects into fingerprint individuality estimates. The <u>model</u> may be too simple to be adequately fit to large databases as demonstrated in Section V.*

To sum up, we can maintain flow in scientific writing either by using linking words as well as phrases, or by repeating key words or phrases.

1.2　Contribution

Contribution（创新）refers to new knowledge obtained in the scientific research. This new knowledge might be either a new method of conducting a research, or a new finding obtained. The new finding might be a new model, a new theory, or a new product, or a new equation. Thus, a scientific research can not be called a research without contribution.

1.3　Documentation

A research paper should create *arguments*（论点）with *evidence*（论据）. One effective way to create or develop arguments is to *document sources*（引证）properly. That is to say, you need to employ accurate citations placed in a correct form so that readers can locate all the sources and verify the data.

Documentation（引证）is indispensable in scientific writing, especially in writing sections of Introduction and Discussion. As a research writer, you need to be able to move back and forth gracefully between conducting your own argument and using material from your research in the form of summary, paraphrase, or quotation.

1) Three common systems

The proper *formats*（格式）for citations and bibliography *entries*（条目）are simply *conventions*（规定）within an academic *discipline*（学科）to facilitate the reader's *retrieval*（检索）of the original sources. As the requirements of journals vary widely from discipline to discipline, and even within the same discipline, it is not possible to offer recommendations that are universally acceptable. In this sub-section, we will present three common systems that are accepted in most disciplines.

(1) Citation order system.

The citation order system is simply a system of citing the references (by number) in the order that they appear in the paper. The number is usually placed in *parenthesis*（圆括号, Please see

E. g. 7) or *squared brackets* (中括号, Please see E. g. 8). This system is quite popular in scientific writing for its advantage like avoiding the substantial printing expense of the name and year system.

E. g. 7: *As it is well-known in the diesel engine community, the process of injecting fuel into the combustion chamber is responsible for the quality of the air/fuel mixture that is achieved (1).*

E. g. 8: *Signal detection, classification, and identification were major challenges in the research and development of radar and wireless communication systems for a greater part of the 20th century [1]. In particular, identification of radar, radios, and various wireless communications became a very important and popular topic around the time of World War II [2]. Indeed, research into radar detection and signature evaluation is ongoing in the military to this day [3], [4]. Transmitter identification has also been of interest to the avionics and cellular industries (in the latter case for the prevention of cellular telephone cloning) [5] - [9]. Outside of military and commercial spheres, since at least the 1990s researchers have explored ways to identify radio transmitters (push-to-talk transmitters, VHF/NHF transmitters, etc.) [10] - [20]. (This paragraph was quoted from Physical-Layer Identification of Wired Ethernet Devices).*

In E. g. 8, we could identify three types of squared brackets. For example, [3], [4] suggest that the idea was borrowed from the third and the fourth entries of the references whereas [5] - [9] reveal that the idea was taken from five entries (from the fifth entry to the ninth entry).

(2) Name and year system.

Name and year system is to place the author's name and the year of publication in parentheses (See E. g. 9 and E. g. 10). The name and year system (often called the Harvard system) has been very popular and is used by many journals and books. Its big advantage is convenience to the author. The disadvantage to the reader occurs when many references must be cited within one sentence or paragraph whereas the disadvantage to the publisher is increased cost (Day and Gastel, 2007).

Name and year system is commonly used in social sciences and is also employed in the journals of *ACM*, *Solar Energy* and *Nuclear Energy*.

E. g. 9: *... with the residential services sector responsible for the 31% and the industrial sector for the 26% (Mirasgedis et al., 2013).*

E. g. 10: *Consumers' choices can define the utility they can derive (Becker, 1965; Lancaster, 1966; Muth, 1966).*

In using the name and year system, you need to deal with the following two situations. In dealing with the page information, you need to include the page number where the information is found if you are quoting a particular passage or citing a particular table. You need to use a comma to separate each element of the citation and use the abbreviation p. (shortened for page) or pp. (shortened for pages) before the page number. In presenting the author's information, you need to place the author's name in parentheses if it is mentioned in a preceding attributive tag. You need to list all three the first time the paper is cited, if the paper has three authors, e. g. "Leech, Jones and Allwright (2014)." If the same paper is cited again, it can be shortened to "Leech et al. (2014)" But when a paper has four or more authors, you need to cite "Leech et al. (2014)" even in the first citation.

(3) Alphabet-number system.

Alphabet-number system is to number the citation order according to the author's first letter of *the last name*(姓) in the references whereas in the *text proper*(正文), the quotation number is out of order, which differs from the citation order system. To be more specific, the cited number in the alphabet-number system is not in numerical sequence as shown in E. g. 11.

E. g. 11: *2D-3D pose tracking or pose estimation is concerned with relating the spatial coordinates of an object in the 3D world (with respect to a calibrated camera) to that of a 2D scene [22], [27]. Although the complete literature review is beyond the scope of this paper, most methodologies can be described as follows: First, one chooses a local geometric descriptor (e. g. points [33], lines [18], [28], or curves [19], [39]) or image intensity [4] that can best quantify features on the image to its corresponding 3D counterpart. Then, explicit point correspondences are established in order to solve the pose transformation. (This paragraph was quoted from "A Nonrigid Kernel-Based Framework for 2D-3D Pose Estimation and 2D Image Segmentation")*

To sum up, we do hope that one of the above systems might help you in your scientific writing. Meanwhile you need to remember that each journal has its own writing formats. Thus as a writer, you have to follow the *Instructions to Authors*(投稿须知) of your *target journal* (投稿的期刊).

2) Rules to follow

To cite the documents in the text, you need to remember the following rules.

(1) List only significant published references.

Considerable research work has been published in each discipline every year, which has made it impossible for you to read all of them. Hence you need to select only relevant literature for your reference. Due to the length requirements, you are required to list or cite only significant published references which are closely related to your research. If you want to cite a relevant paper that has been accepted for publication, you can cite the name of the journal followed by "*In press*" (已录用) or "Forthcoming" in your references.

(2) Check all parts of every reference against the original publication.

To avoid making mistakes in your references and help the readers locate the sources, you need to check all parts of every reference against the original publication before you submit your manuscript. As Day and Gastel suggest that "Make sure that all references cited in the text are indeed listed in the Literature Cited and that all the references listed in the Literature Cited are indeed cited somewhere in the text" (2007:76).

1.4 Language Use

Scientific writing is required to be accurate, brief and clear. To reach this aim, you are supposed to notice its style and flow in the process of scientific writing. Scientific writers need to be sure that their communications are written in the appropriate style. This appropriate style involves the vocabulary and grammar as Swales and Feak suggested in 2010.

1) **Vocabulary**

The first distinctive feature of scientific writing style is to choose the more formal alternative when a verb, a noun, or other *parts of speech* (词性) are selected. The choice is often between spoken English (informal) and written English (formal). For example, the verb "get" is less formal than "obtain" whereas the adjective "numerous" is more formal than "many." Another choice is often made between a phrasal or prepositional verb (verb + preposition) and a single verb or a noun. For example, in academic English, "make use of" will be substituted by "employ"; "get rid of" will be less formal than the verbs "eradicate" and "delete."

2) **Grammar**

To be formal in grammar, you are required to avoid *contractions* (缩写) like "don't," "won't," and "doesn't" and use more appropriate formal negative forms such as "no," "little" and "few" in your scientific writing. Meanwhile, you had better avoid the use of direct questions. For example, "What can be done to solve the problem?" should be replaced by "We need to consider what can be done to solve the problem, or we need to consider how the problem may be solved." In addition, you should avoid addressing the reader as "you." Thus in the following situation, a passive voice is preferred "The results can be seen in Table 1" to "You can see the results in Table 1." Finally remember to place adverbs within the verb.

1.5 Tips on Effective Scientific Writing

1) **Write accurately**

Writing accurately is required for scientific English writing, but grammatical errors are common among Chinese writers. *The most common errors are agreements between subjects and verbs and between singular and plural nouns due to negative transfer of Chinese mother tongue* (由于母语的干扰,最常见的错误是主谓不一致以及名词的单复数) which could be avoided if you develop your awareness. Other common errors are misuse of words and lack of linking words.

E. g. 12: *Due to the work piece can not be absolute cylinder, the work piece must produce axial movement leads to a decrease in the quality of welding.*

In E. g. 12, the first error is that "due to" has been followed by a sentence and the main clause has two verbs (must produce and leads to). As "due to" is a prepositional phrase that can be followed by a noun or a noun phrase, "due to" can be changed into "because" whereas in the main clause the linking word "which" should be added after "the work piece."

Another **common error** is Chinese English (*Chinglish* 中式英语) that often occurs when the Chinese writer translates the ideas into English in the process of scientific writing because the writer has little knowledge of the *idiomatic expressions* (习惯用语) and collocations. For example, "The researches for the micro ball end mills are not a lot" was a word for word translation, as the writer has no idea of the idiomatic expression "Few researches have been done on the micro ball end mills." Another similar example is "Currently, influences of elastic characteristics have been researched deeply." This sentence can be improved as follows "Currently, numerous researches have been done on the influences of elastic characteristics. " To write more idiomatically in scientific English writing, we suggest that you read more research papers written by native writers.

As collocations play an important role in expressing accurately, the Chinese writers are required to remember as many English collocations as possible. The following example shows us a wrong collocation"... finally gives a search recommendation model based on the similarity between each two users," in which "gives" should be changed into "establishes. " More collocations should be remembered. Namely, the word "same" should be collocated with "the" (the same).

2) Write briefly

Writing briefly is also required in scientific writing, in which unnecessary words or phrases and even sentences should be deleted. That is to say, you can simplify your expressions by reducing or removing unnecessary words. Now look at the following two examples.

E. g. 13: *It is easy to see that all of the title length is between 5 words and 12 words.*

Improved:

The titles' length ranges from 5 to 12 words.

The improved sentence is brief and clear with nine words whereas the original sentence with 18 words is complex and difficult for the reader to grasp the main idea.

E. g. 14: *Out of sixteen researching articles, eight of them are from Elsevier and the rest are from AIAA and Acta Astronautica. Actually, five of them are from AIAA and three of them come from Acta Astronautica. (33 words)*

Improved:

Out of sixteen researching articles, eight of them are from Elsevier, five from AIAA and three from Acta Astronautica. (19 words)

In E. g. 14, two sentences have been employed correctly in terms of grammar in the original sentence, but they are not as clear and brief as the improved sentence. This improved sentence has successfully employed a parallel structure to express the idea clearly and briefly.

The above two improved sentences are much shorter and clearer than their original versions, suggesting that you can simplify your expressions to make the ideas more effective.

Another way to simplify your expressions is to reduce the *restrictive relative* (限制性定语从句) if ① the relative clause consists of a passive verb plus some additional information; ② the relative clause contains the relative pronoun, an adjective ending in -ble, plus additional information; ③ the relative clause contains the verb *have*. In this case the relative pronoun and *have* can both be dropped and replaced by with.

E. g. 15: *A robot is a multi-programmable device, which is capable of performing the work of a human.*

Improved:

A robot is a multi-programmable device capable of performing the work of a human.

E. g. 16: *The cell is covered by an aluminum lid to create a closed environment within the cell, which has the exception of the inlet and outlet gas.*

Improved:

The cell is covered by an aluminum lid to create a closed environment within the cell, with the exception of the inlet and outlet gas.

3) Write clearly

To write clearly refers to clear organization and clear vocabulary. As we mentioned clear organization in **1.1,** we will focus on clear vocabulary.

E. g. 17: *Many students can not express their ideas clearly in their writing and speaking. One of the reasons may be the vocabulary.*

Improved:

Many students can not express their ideas clearly in their writing and speaking. One of the reasons may be their insufficient vocabulary.

E. g. 17 failed in expressing clearly due to the use of the word vocabulary, which requires the word insufficient to be added.

4) Avoid long sentences

Brittman (2009:4) found that "very long sentences are especially common in Chinese-English writing because the writers often translate directly from Chinese to English. Although, in Chinese writing it is acceptable to put several supporting ideas in one sentence to show their relationship, in English, the main idea and each supporting idea are typically written in separate sentences." To illustrate his claim, we provide two example sentences written by two of our students.

E. g. 18: *One approach to actuator location optimization is to consider open-loop measures and the location is typically chosen so as to enhance the controll ability properties of the Actuator.*

Improved:

One approach to actuator location optimization is to consider open-loop measures. In this approach, the location is typically chosen so as to enhance the controll ability properties of the Actuator.

In E. g. 18 the improved sentence is much clearer when the original sentence is separated into two.

E. g. 19: *This will guarantee that the controlled system is a Hilbert-Schmidt operator, and also that the optimal actuator problem is well-posed, which can be weakened for certain classes of systems, and there are generally a finite number of disturbances, control signals, and outputs.*

Improved:

This will guarantee that the controlled system is a Hilbert-Schmidt operator, and also that the optimal actuator problem is well-posed. These assumptions can be weakened for certain classes of systems. However, in practice, there are generally a finite number of disturbances, control signals, and outputs.

In E. g. 19 the original sentence is too long for the reader to grasp the major point. But it is more comprehensible when this sentence is separated into three sentences.

5) Place the most important subject at the beginning of the sentence for emphasis

As Brittman stated that "Chinese writers often preface the main topic of a sentence by first stating the purpose, location, reason, examples and conditions as introductory elements. However, this has the effect of demoting the importance of the main idea and making the reader think the author is indirect. Bring the main idea to the beginning of the sentence stating any locations, reasons, etc., afterwards" (2009:6). The following three examples might be adopted for your reference.

E. g. 20: *Based on the conventional methods, the anti-penetration mechanism is expected to be characterized quantitatively.*

Improved:

The anti-penetration mechanism is expected to be characterized quantitatively based on the conventional methods.

E. g. 21: *Unfortunately, although the literature has reported how to synthesize cube structured materials, most of synthesis technologies are sophisticated, time-consuming.*

Improved:

Unfortunately, most of synthesis technologies are sophisticated, and time-consuming, though the literature has reported how to synthesize cube structured materials.

E. g. 22: *Especially when numerical control (NC) techniques[4] are widely used in industry and rapid prototype methods[5][6] bring huge economical benefits, the advantage of constructing 3D model[7][8][9] becomes extremely obvious.*

Improved:

The advantage of constructing a 3D model[7][8][9] becomes extremely obvious especially when numerical control (NC) techniques[4] are widely used in industry and rapid prototype methods[5][6] bring huge economical benefits.

The above three examples failed in placing the most important subject at the beginning of the sentence for emphasis, which is quite common among Chinese writers. Thus, this type of error should be avoided in scientific writing.

6) Place the old information before the new

The order of information might make the ideas either clear or incoherent, which should draw enough attention from Chinese writers. The following example is a good case.

E. g. 23: *For comparison, the following leader selection algorithms are simulated. A random subset of agents is chosen to act as leaders in the first algorithm.*

Improved:

For comparison, the following leader selection algorithms are simulated. In the first algorithm, a random subset of agents is chosen to act as leaders.

In the second sentence of E. g. 23, "a random subset of agents" is new information whereas "in the first algorithm" is old information. Hence the improved sentence is more effectively expressed when the order of information has been changed.

1.6 Summary

In this chapter, we have provided an overview of some important features of scientific writing such as clarity, contribution, documentation and language use with examples. In addition, we have offered tips for effective writing via example analyses. Hence we do hope that with more practice you will improve your writing performance.

Tasks

Task 1
Directions: Identify the basic components while reading a paper in your field. Write down the basic components each time you finish reading a journal paper. Check whether there are any similarities and differences after you read five papers from different journals.

Task 2
Directions: Mark the linking devices when you are reading your research paper. Write down different types of linking devices in your notebook and use them in your future writing.

Chapter 2
How to Write a Title

A title is important in scientific papers. It is like the brand of the paper, with only a few of words to signify the contents of the paper and the *essence*(精华) of the research. Because it is like the brand of the paper, a good title should attract readers that there is something worthy to be read and to inform readers what the papers are about.

As a brand of the paper, the title is the most visible part. It is the part that will be read most frequently and sometimes it remains the only part of the paper that will be read at all. Mabe and Amin (2002) found, for example, that 5,000 science readers estimated that they read, on average, 97 articles per year, twice as many (204) abstracts, and ten times as many (1,142) titles. To increase the possibility of being read, a title must satisfy the requirements to address the proper audience, and an improperly titled paper will definitely cause the loss of its potential readers. So writing a good title is a challenging process, which takes time and imagination.

2.1 Writing Principles: BAD Principles

Day and Gastel (2007) defined that a good title should use the fewest possible words to adequately describe the contents of the paper. Baker (2011) claimed that the best title should *detail* (详述) the major researching results performed. So writing a title for a scientific paper is not easy, which requires writers to consider how to represent the essence of the research, how to detail the major findings of the work, and how to condense them in only a few of words. From this aspect, an attractive title can not be long, general and *random*(任意的) to challenge the readers' patience. It is better presented in such brief, accurate and specific words that can easily distinguish themselves from others in a similar field.

1) Brevity

Ideally, a well-written title should be brief and *concise*(简洁). It should be short and effective, but not too short. Some researchers analyzed the title length to try to get the exact numbers to present a good title. Hu (2000) suggested that generally a title is composed of no more than 20 words. Anthony (2001) researched 600 paper titles from journals of IEEE computer society and found that the average length of a title is about 8 to 9 words. Soler (2007) analyzed 240 research paper titles from journals of *Natural Science on Biology*, *Medicine and Boichemistry*. The findings showed that the average title length is about 14-15 words. Though the title length has *field variation*(专业差别) according to the researchers, titles in scientific journals seem to have been getting longer and thus more *informative*（提供信息的）(Hartley, 2005), but not *rambling*（漫无边际的）.

It is true that the more informative titles are, the more effectively they serve their functions. But it does not mean that the more informative titles are, the more words they have. A good title still should be the contents *condensed*（浓缩的）, to represent and signify the essence of the research, and can not be too difficult to understand. This requires using words that are *representative*（有代表性的）and *essential*（必要的）to the paper, and some words like "cool" *jargons*（行话）, *slangs*（俚语）and *abbreviations*（缩写）should be avoided in titles.

Metaphors（比喻）and creative writing can be fun in titles of opinion papers, but in general they should be *left out of*（省去）research titles. Few readers will search for metaphorical words, so use words that make direct and immediate sense to readers.

Usually, the excessively long titles contain some *misleading*（令人误解的）and useless words, which often occur at the right side of the title, like the unnecessary articles, "A," "An" and "The"; some noun phrases denote nothing just taking up space like "Studies on," "Investigations on," "Observations on," "Research on," "Analysis of," "Development of," "Report of," "Review of," "Treatment of," and "Use." The best choice is to omit them for brevity.

2) Accuracy

To pursue brevity does not suggest *confusion*（混淆）and *ambiguities*（含糊不清）in title design. A well-written title should be accurate, clear and exact, to summarize the research completely, to *depict*（描述）the design and findings exactly without any misunderstandings. For example, in the paper *Dictyostatin Flexibility Bridges Conformations in Solution and in the β-Tubulin Taxane Binding Site* (Joqulekar et al., 2011), the complex, *flexible polyketide macrolide-dictyostatin* was mainly employed under the situation of extensive force-field-based *conformational* searches, and the results display a diversity of *conformations for dictyostatin* to illustrate the *flexibility*. One important conformation is a *binding-site interaction* between *DCT and assembled materials in solution*. According to the paper, the title *Dictyostatin Flexibility Bridges Conformations in Solution and in the β-Tubulin Taxane Binding Site* has accurately and completely summarized the purpose and important findings of the research, that is *dictyostatin flexibily bridges comformations*, and the research method has been depicted exactly as well, i. e *in solution (water) and B-Tubulin Taxane Binding Site*. So this is a well-written title for its brevity and accuracy.

In creating titles, *syntax*（句法）should be especially careful, because most grammatical errors are often made by the faulty word order, which misleads readers.

A paper studied the transesterification of glycerol trioleate over a basic ionic liquids, the product of 1-butyl-3-methylimidazolium hydroxide and meghyl ester is achieved. But the title *Biodiesel preparation catalyzed by basic ionic liquids from transesterification of glycerol trioleat* really misleads readers. If the title can be *Biodiesel preparation from transesterification of glycerol trioleate catalyzed by basic ionic liquids*, the relations between *biodiesel preparation* and *basic ionic liquids* will be clearer, and the title will be more accurate to reflect the essence of the research.

3) Distinctiveness

For a paper, a brief, accurate title is not enough. It is still supposed to be specific and *distinctive*(区别性) to distinguish themselves from other treatments of the same general subjects. To create a distinguished title, the research design *identification*(识别) is crucial. It enables retrieval for a specific type of evidence; readers may then quickly judge whether the paper is relevant. For example, if a reader is specifically looking for the effect of *Baogan Yihao*, for instance, as the research question is about the efficacy of intervention, it is likely that in a list of many potentially relevant papers, he or she will select the paper that is clearly distinguished as *anti-fibrosis of Baogan Yihao* in the title.

In creating the title for the research, only listing *Effect of Baogan Yihao on Liver Fibrosis* is not specific enough to distinguish itself from other researches on *Baogan Yihao*, because readers can not know whether *Baogan Yihao* enhances or prevents *the process of liver firbosis*, whether this effect can be applied in many groups, for instance, human beings. So the design, the method, the findings, etc. are necessary to be considered. If the word "effect" can be changed to the findings *anti-fibrosis*, and *liver fibrosis* can be modified with the cause $CC1_4$ *and High-Fat Feeding in Rats*, the meaning might be clearer, like the paper with the title *Baogan Yihao protection on liver fibrosis induced by $CC1_4$ and high-fat feeding in rats.*

On the basis of the BAD principles—brevity, to use accuracy and distinctiveness, a good title, therefore, can be defined to use as the **fewest** possible words that **accurately** describe the **specific** contents of the paper.

2.2 Common Features

Titles are labels, which are mostly composed of noun phrases. Li (2012) analyzed 4,492 scientific paper titles from 50 subjects. The result showed that noun and noun phrases are the most frequent syntax structures (84.5%) followed by *gerund*(动名词) titles (4.5%) and *colonic*(冒号的) titles (1.2%). And adjectives rarely occur in titles, including "new," "novel" and "efficient" (Yu, 2006).

1) Noun+prepositions

In nouns and noun phrases structures, *prepositions*(介词) and *articles*(冠词) are the most frequently used words (Anthony, 2001: 194). Among the noun structures, noun+for/of is the top (Wang, 2011; Ma, 2014).

In noun+for structure, the most common way is to describe how or in what context a newly developed technique or strategy can be used. For example, *Designing Access Methods for Bitemporal Databases.* Another use of noun+for structure is to narrow the focus of the research, for example, *A Priority-Driven Flow Control Mechanism for Real-Time Traffic in Multiprocessor Networks.*

In noun+of structure, relationship, *properties*(属性), characteristics, etc. are often followed by "of." The nouns often going with "of" are often *abstract*(抽象的) nouns, like:

(1) To show the substance performance and characteristics: characterization of, mechanism of, stability of, stabilization of, properties of, etc.

(2) To show the research design: synthesis of, detection of, assembly of, polymerization of,

comparison of, preparation of, application of, investigation of, etc.

(3) To show the related effect of research objects: effect of, hydrolysis of, identification of, influence of, quenching of, solubilization of, etc.

(4) To show the changes of substance: formation of, imaging of, sensing of, reactivity of, step of, etc.

Besides the above two types of *nominal constructions*(名词结构), noun/noun phrases+to often goes with *post-modifiers*(后置定语), and noun/noun phrases mostly consist(s) of the word "approach" or "application," like *A Dissipativity Approach to Safety Verification for Interconnected Systems* or *On Set-Valued Kalman Filtering and Its Application to Event-Based State Estimation*. Sometimes, noun/noun phrases+to is used to indicate the purpose of the research in the title, like *One-Step Inversion Process to a Janus Emulsion with Two Mutually Insoluble Oils*.

2) Gerund titles

Gerunds are verb forms ending in-*ing* and considered as nouns in sentences or phrase structures. In titles, gerunds (e.g., improving or expanding or using) imply some sort of action without identifying the actor. Like in the title *Incorporating a built environment module into an accelerated second-degree community health nursing course*, the gerund *incorporating* suggests the action of making a built environment module part of health nursing course. As for the actor of incorporation, it is not necessary to give details in the title. According to the research (Ma, 2014), the word *using* seems the most popular one in titles to introduce the method of the study.

Though Hu (2000) suggested that gerunds are also common noun structures occurring in titles, to some editors (Williams, 2009; Hays, 2010), they decried most nominalizations of verbs. Because they believed that neither the actor nor the action was a key concept or variable in the research, so the gerund was *superfluous*(多余的) and did no important work in titles. As for the readers of this book, our suggestion is to look before you leap in using gerunds.

3) Incomplete sentences

As we discussed at the beginning, a title is like the brand of the paper, using the brief and economic words to summarize the essence of the research. Therefore, incomplete sentences with nominal group constructions are more common in titles. For example, *A Priority-Driven Flow Control Mechanism for Real-Time Traffic in Multiprocessor Networks*, it is a noun phrase with a preposition *for*. As for *A Dissipativity Approach to Safety Verification for Interconnected Systems*, it is a noun phrase with *to* structure to modify the previous noun. And in *Incorporating a built environment module into an accelerated second-degree community health nursing course*, a nominal group construction consisting of the gerund *incorporating* makes up the whole title.

While, there is a new tendency recently, *full-sentence construction*(完整句子结构) is becoming more and more common with a clear statement and an unambiguous conclusion. Comparing with the titles in phrases, sentences in statement sometimes may give more *explicit*(清楚的) information about the result of the study, like the title *Dictyostatin Flexibility Bridges Conformations in Solution and in the β-Tubulin Taxane Binding Site*. But right now, the full-sentence title is more field specific in biology research papers (Soler, 2007).

4) Colonic titles

Though most titles are not sentences, punctuations are still common. The highest frequently used one is colon. The titles with colons are called colonic titles or hanging titles (there is rarely a dash connecting in hanging titles). Comparing with science of engineering, colonic titles are commonly used in papers of science of arts, like business, economics and applied linguistics (Fontanet *et al.*, 1997).

The common ways to structure colonic titles are the Name: Description relationship. For example, *MPS: Miss Path Scheduling for Multiple-issue Processors*, (Sanjeev *et al.*, 1998) or Topic: Scope relationship, for example, *Collaborative Multimedia Systems: Synthesis of Media Objects* (Candan *et al.*, 1998). But there is still a wide variation in the use of these two forms across different research fields.

2.3 Specific Analyses

As mentioned above, a well-written title should be brief, accurate and distinctive, therefore in creating a title, the *intelligibility*(完整性) of these principles but any individual of them needs considering.

E. g. 1: *Effects of pulsed magnetic field on the micro-hardness of HSS cutting tool materials.*

This is an *indicative*(陈述的) title (*with* 13 *words*), because the purpose rather than the conclusion of the research has been mentioned. This title informs us that the paper is about the association between pulsed magnetic field and micro-hardness of HSS cutting tool materials (research question). But we do not know if the pulsed magnetic field was found to be effective or not, and the title does not give any *indication*(迹象) regarding the study design. So according to the BAD principles, this title is not **distinctive**.

From the paper, it is found that the pulse magnetization parameter can obviously improve the specimen micro-hardness of HSS material. The word "effect" in the title could, therefore, be improved to become more informative.

Improved: Pulsed magnetic fields enhance the micro-hardness of HSS cutting tool materials.

E. g. 2: *The applications and design methods of acoustic metamaterials about passive density and modulus.*

This is still an indicative title with 13 words. This title tells us that the research is about the physical characteristics (passive density and modulus) of acoustic metamaterials. Though the purpose and method of the research have been listed, they are not specified, so this title is not **brief** and **distinctive**. According to the paper, the passive density and modulus of acoustic metamaterials were analyzed in FEM (Finite Element Method), the result showed that the metamaterails offer a great choice on material selection for wave manipulations. As for the applications, it seems not the focus of the research. So this title may be:

Improved: Passive density and modulus of acoustic metamaterials in finite element method.

E. g. 3 *Hyperspectral Imaging System for Plant Disease Diagnosis with LCTF.*

This is an indicative title with 9 words. From the title, it can be seen that the paper discussed the method of plant disease diagnosis. But it is not clear whether the process of disease diagnosis carried out by liquid crystal tunable filter (LCTF) or the hyperspectral imaging system employs liquid crystal tunable filter (LCTF). So this title is not **accurate** for the misplacement of the words "with LCTF." Meanwhile, the abbreviation is not suggested in the title. According to the result, it indicated that to identify and classify vegetable diseases using hyperspectral imaging technology was feasible. Therefore, the title can be:

Improved: Hyperspectral imaging system with liquid crystal tunable filter for plant disease diagnosis.

E. g. 4: *Electro-driven system design in Formula Student Racing Cars.*

This is still an indicative title with 8 words. The title informs(告诉) us that the paper is about a new design of formula student racing cars (research question), but it does not present the specific method of the design and the result of the new design, so it is not **distinctive** and **accurate**. From the paper, it is known that the formula student racing car has been optimized in the motor, controller, gearbox, and batteries, and the power matching is designed as well. The test results prove the design requirements are obtained.

Improved: Optimizing control factors in electro-driven system enhances Formula Student Racing Cars.

E. g. 5: *Effect of Baogan Yihao on Liver Fibrosis in Rats Induced by CCl_4 and High-Fat Feeding.*

This indicative title is 15-word long. It lacks information on the result of the study, and the past participle "induced" brings ambiguities to the object of the research. So the title is not **accurate** and **distinctive**. From the paper, it showed that this study focused on liver fibrosis which was resulted by CC14 and high-fat feeding. The experimental result showed that Baogan Yihao can reduce the liver pathological changes. So if the "effect of" can be modified like "positive effect of," it is more acceptable, but still not especially good because the result of the research is too general.

Improved: Baogan Yihao reduces liver fibrosis induced by CCl_4 and high-fat feeding in rats.

E. g. 6: *A new method to evaluate the combustion chamber profile of diesel engine pistons in an energy spectrum perspective.*

This title is informative with 18 words long. It shows the purpose and method of the research. From the title, it can be seen that the research is about the evaluation of the combustion chamber profile of diesel engine pistons. And the result showed that the combustion chamber profile could be evaluated properly with the use of the energy spectrum perspective. So the energy spectrum is the contribution of the paper. But it is not emphasized as it should be, because the focus has been postponed to the end of the title. Therefore this title is not **brief** and **distinctive**. Suppose the focus can be put at the beginning of the title, then things will be better:

Improved: Energy spectrum in evaluating the combustion chamber profile of diesel engine pistons.

2.4 Summary

It is true that with BAD principles, writing a good title is still not easy. It is the process of suffering, involving brainstorming, constructing, creating, revising, and even refusing. So the best solution is, under the supervision of BAD principles, to practice, and practice, because practice makes perfect.

Tasks

Task 1
Directions: Analyze the titles in your research field and identify their common features.

Task 2
Directions: Compare your analyses of the title with those of your partner and see whether there are any similarities and differences.

Chapter 3
How to Write Abstracts

In the "old days," most papers did not have abstracts; surprisingly perhaps, abstracts were only introduced into medical research articles during the 1960s (Swales and Feak, 2009: 1). Now most research articles (RA) are structured as follows: Abstract, Introduction, (Materials and) Methods, Results and Discussions/Conclusions (often shortened as AIMRD/C).

When writing an abstract, we need to consider journal-specific requirements as well as disciplinary differences, therefore, it is very necessary to refer to **Instructions for/Notice to Authors** of a target journal. Although the style of an abstract may differ from discipline to discipline and from journal to journal, the structure and information provided are quite similar. The aim is always to tell readers all they need to know to help them decide whether to buy/read the paper (Wallwork, 2011:180).

In this chapter, we will mainly deal with the *basic components* (基本要素), the linguistic features of RA abstracts, the tips on writing good RA abstracts and the *syntactic templates* (句式) respectively.

3.1 Basic Components

In terms of the contents, there are basically two major types of RA abstracts owing to the different requirements of different disciplines and journals. One is a brief summary of the main sections of the paper (Day, 1988:52; Swales, 2001:211), while the other is a "result-driven" abstract which focuses primarily on one or two aspects of the study, usually — but not always — the methods and the results (Swales, 2001:211; Glasman-Deal, 2011:199).

The summary type can be further divided into two subtypes depending on whether there is the background section required in the specific journal. One is a four-move structure (IMRD/C) which briefly summarizes each of the main sections of the paper: Introduction, (Materials and) Methods, Results and Discussions/Conclusions. One representative of this structure is Day (1988) who stated "The Abstract should: ① state the principal objectives and scope of the investigation; ② describe the methods employed; ③ summarize the results, and ④ state the principal conclusions. (52-53)." The other is a five-move structure (BPMRC) as mentioned by Weissberg and Buker (1990):

| Some background information | B |
| The principal activity or purpose of the study and its scope | P |

Some information about the methods used in the study	M
The most important results of the study	R
A statement of conclusion or recommendation	C

(Cited from Cargill and O'Connor, 2013:70)

Although they use different terms, yet the only difference between these two subtypes is whether there is the background section in the abstract which is usually separated or marked by words or phrases like "this paper presents...," "in this paper/study, we propose...," and "... is presented/proposed in this paper".

However, in terms of the structure of the summary type, most journals require the *unstructured abstracts*(非结构式摘要)in only one paragraph, while some journals (the *Medicinal Chemistry* for example) require *structured abstracts* (结构式摘要) in several paragraphs with subheadings. Actually, the structured abstracts are just unstructured abstracts with subheadings. Therefore, in 3.2, five examples will be given to illustrate five types of abstracts, namely the unstructured BPMRC, the structured BPMRC, the unstructured IMRD/C, the structured IMRD/C and the result-driven type.

3.2 Specific Analyses

1) The summary type
(1) The unstructured BPMRC abstract (Table 3.1).

Table 3.1 Component analyses of Abstract 1

S	Original Sentences	Component Analyses
S1	Road traffic sensors provide rich multivariable datastreams about the current traffic conditions.	These three sentences introduce the research background and the research gap by using "... provide rich...," "there are unusual..." and "... is useful to..."
S2	Occasionally, there are unusual traffic events (such as accidents, jams, and severe weather) that disrupt the expected road traffic conditions.	
S3	Detecting the occurrence of such events in an online and real-time manner is useful to drivers in planning their routes and in the management of the transportation infrastructure.	
S4	We propose a new method for detecting traffic events that impact road traffic conditions by extending the Bayesian robust principal component analysis (RPCA) approach.	This sentence introduces the research problem and the method by using "We propose a new method... by extending...approach."

continued

S	Original Sentences	Component Analyses
S5	Our method couples multiple traffic datastreams so that they share a certain sparse structure.	These five sentences describe in detail the research method by using "Our method couples...," "...is used to...," "...are measurements of...," "Our proposed method processes...," "We experimentally analyze... using..."
S6	This sparse structure is used to localize traffic events in space and time.	
S7	The traffic datastreams are measurements of different physical quantities (e.g. traffic flow and road occupancy) by different nearby sensors.	
S8	Our proposed method processes datastreams in an incremental way with small computational cost; hence, it is suitable to detect events in an online and real-time manner.	
S9	We experimentally analyze the detection performance of the proposed coupled Bayesian RPCA (BRPCA) using real data from loop detectors on the Minnesota I-494.	
S10	We find that our method significantly improves the detection accuracy when compared with the traditional PCA and noncoupled BRPCA.	This sentence indicates the contributions of the research by using "We find that our method significantly improves... when compared with the traditional..."

Cited from YANG, *et. al.*, 2014:1936.

Structurally, this abstract follows the BPMRC model. The ten sentences present logically and smoothly the research background (Sentences 1-3); the research problem and the method (Sentence 4); the methods in detail (Sentences 5-9) and the result and the implication of the research (Sentence 10). Obviously, the methods section has been emphasized by five sentences.

Language focus:

Just as there are variations in the structure and the contents of RA abstracts, there are also variations in the linguistic features, such as the *tenses* (时态), the *voices* (语态) and the *statement styles* (有人称句或无人称句). Therefore, when we write an abstract, it is also necessary to check the linguistic features before we submit a paper to a target journal.

In terms of these features, there are mainly two contradictory viewpoints. One maintains that "The abstract is characterized by the use of **past tense, third person, passive**, and **the non-use of negatives.**" (Graetz, 1985. Cited from Swales, 1990:179). However, the other proposes that "over the last decade or so, academic writing **has gradually lost its traditional tag as an**

objective, faceless and impersonal form of discourse and come to be seen as a **persuasive endeavour involving interaction between the writer and the readers**" (Hyland, 2008:65). In line with the changing features in the full scientific article, it is possible that front matter (**the foregrounding of results, the use of the present tense in the active form and the apparent manipulation of the basic IMRC/D structure**) is being included in journal abstracts at the expense of their summarizing content (Swales, 1990:180).

Table 3.2 presents the linguistic features of Abstract 1. Please note: only the main verbs have been analyzed.

Table 3.2 Linguistic features of Abstract 1

Sentences	Sections	Tenses	Voices	Styles
S1	Background	simple present	active	impersonal
S2	Background	simple present	active	impersonal
S3	Background	simple present	active	impersonal
S4	Introduction+Method	simple present	active	personal
S5	Method	simple present	active	impersonal
S6	Method	simple present	passive	impersonal
S7	Method	simple present	active	impersonal
S8	Method	simple present	active	impersonal
S9	Method	simple present	active	personal
S10	Results and Conclusions	simple present	active	personal

Linguistically, besides using the simple present tense and the active voice, the authors also choose to use first person pronouns "We propose" in Sentence 4, "We analyze" in Sentence 9 and "We find..." in Sentence 10 and "Our method" (Sentence 5) and "Our proposed method" (Sentence 8) to foreground their contributions. Hyland (2002) found that writers choose to announce their presence where they make a knowledge claim. "At these points, they are best able to explicitly foreground their distinctive contribution and commitment to a position." (Cited from Livnat, 2012: 93-94)

(2) The structured BPMRC abstract.

The now-fashionable so-called "structured" abstract (i.e. with named subsections) did not appear until about 1987 (Swales and Feak, 2009: 170). For example, ***Information for Authors*** of *Annals of Botany* requires that "The second page should contain a structured Abstract not exceeding

300 words made up of bulleted headings." For "ORIGINAL ARTICLES," these headings will normally be as follows:

- Background and Aims
- Methods
- Key Results
- Conclusions

The following example is chosen from that journal (Table 3.3).

Table 3.3 Component analyses of Abstract 2

Sections	Sentences	Original Sentences	Component Analyses
Background and Aims	S1	AtSUC2 encodes a sucrose/proton symporter (蔗糖/质子协同载体) that localizes throughout the collection and transport phloem and is necessary for efficient transport of sucrose from source to sink tissues in Arabidopsis thaliana (拟南芥).	These two sentences introduce the Background and Aims of the research by using "... is necessary for efficient transport of...," "... have severely delayed development and stunted growth and, in previous studies, failed to complete... by producing viable seed".
	S2	Plants harbouring homozygous(纯合子的;同型结合的) AtSUC2 null alleles (无效等位基因) accumulate sugar, starch, and anthocyanin (花色素苷,花青素,花色苷) in mature leaves, have severely delayed development and stunted growth and, in previous studies, failed to complete their life cycle by producing viable seed.	
Methods	S3	An AtSUC2 allele with a T-DNA insertion in the second intron (基因内区;内含子) was analysed.	These three sentences describe the Methods by "... analysed," "... is not produced," "... did not catalyse...," "... were grown..., and... recorded."
	S4	Full-length transcript from this allele is not produced, and a truncated(切去顶端的,平头的) protein translated from sequences upstream of the insertion site did not catalyse sucrose uptake into yeast (酵母,酵母菌), supporting the contention (争辩;论点) that this is a null allele.	
	S5	Mutant plants were grown in a growth chamber with a diurnal (昼行性的,白昼的) light/dark cycle, and growth patterns recorded.	

Sections	Sentences	Original Sentences	Component Analyses
Key Results	S6	This allele (SALK_038124, designated AtSUC2-4) has the hallmarks(标志,印记) of previously described null alleles but, despite compromised carbon partitioning and growth, produces viable seeds.	These two sentences summarize the Key Results by using "…produces viable seeds" and "…was chronologically delayed but occurred at the same point.…"
	S7	The onset of flowering was chronologically delayed but occurred at the same point in the plastochron(叶原基间隔期,间隔期) index as wild type.	
Conclusions	S8	AtSUC2 is important for phloem([植]韧皮部) loading and is therefore fundamental to phloem transport and plant productivity, but plants can complete their life cycle and produce viable seed in its absence.	These two sentences emphasize the Conclusions by using "…is important for…is therefore fundamental to…, but plants can complete their life cycle and produce viable seed…," "…appears to have mechanisms for mobilizing…into developing seeds independent of…"
	S9	Arabidopsis appears to have mechanisms for mobilizing reduced carbon from the phloem into developing seeds independent of AtSUC2.	

Cited from Srivastava C., et. al., 2009: 1121.

Structurally, this abstract is clearly structured by the subheadings (named subsections): **Background and Aims, Methods, Key Results and Conclusions.** Each section of the paper is evenly summarized by two or three sentences as Sentences 1 and 2 introduce the Background and Aims, Sentences 3, 4 and 5 describe the Methods, Sentences 6 and 7 summarize the Results, and Sentences 8 and 9 emphasize the Conclusions.

Linguistically, as shown in Table 3.4, the Background and Aims section (Sentences 1-2) and the Conclusions section (Sentences 8-9) are mainly constructed in the simple present tense and active voice as they are established truth or fact and the possible implications, while the past passive is used mainly in the Methods section (Sentences 3 and 5) and the Results section (Sentence 7) to refer to the specific study carried out in the past. What's more, all the sentences begin with the impersonal subject as the authors want to emphasize the objectivity and impersonality of the research.

Table 3.4 Linguistic features of Abstract 2

Sentences	Sections	Tenses	Voices	Styles
S1	Background and aims	simple present	active	impersonal

continued

Sentences	Sections	Tenses	Voices	Styles
S2	Background and aims	present perfect	active	impersonal
S3	Methods	simple past	passive	impersonal
S4	Methods	simple present + simple past	active + passive	impersonal
S5	Methods	simple past	passive	impersonal
S6	Results	simple present	active	impersonal
S7	Results	past present	passive	impersonal
S8	Conclusions	simple present	active	impersonal
S9	Conclusions	simple present	active	impersonal

(3) **The unstructured IMRD/C abstract.**

In some journals, the Background section is optional, therefore, the IMRD/C model is very common (Table 3.5).

Table 3.5 Component analyses of Abstract 3

S	Original Sentences	Component Analyses
S1	An efficient single-frequency Nd：YVO4 master-oscillator(振荡器) power-amplifier (放大器) is described.	This sentence introduces the research problem by using "An efficient... is described."
S2	Gated(门控,门限) pulses from a CW diode-pumped ring laser were amplified by three Nd：YVO4 amplifiers.	These three sentences describe the methods by using "... were amplified by...," "... were obtained..." and "... were used to..."
S3	Pulses of 1 mJ energy; 2 kW peak power and 1 ls duration were obtained at a repetition frequency of 10 kHz.	
S4	Analogue shaping of the input pulses was used to control the output pulse shape and thus extend the useful range of pulse widths from the 100 ns to the 1s regime.	
S5	A simple mathematical expression was used to model the results which provided reasonable agreement with the experimental work.	This sentence describes both the method and the result by using "a simple... was used to model the results which provided reasonable agreement with the experimental work."

S	Original Sentences	Component Analyses
S6	This system offers a promising pump source for a quasi-CW singly-resonant optical parametric oscillators.	The sentence indicates the implication of the research by using "This system offers a promising pump source for…"
Cited from Yarrow, *et al.*, 2007: 361.		

Structurally, it's the IMRD/C model. There are altogether six sentences. The research problem is introduced by Sentence 1. Sentences 2, 3 and 4 describe the methods. Sentence 5 summarizes both the method and the result. And the implication is summarized in Sentence 6. Obviously, the Methods section is emphasized and explained in detail by three sentences.

Linguistically, one noticeable feature of this abstract is the *parallelism* (排比) established by the impersonal subject and the past passive form of the main verbs. Especially, the past passive forms of "…were amplified…," "…were obtained…," "…were used to control…" and "…were used to model…" are used to describe the research method objectively and impersonally. What's more, the passive construction helps the authors to foreground their contributions of the research by putting "An **efficient** single-frequency Nd:YVO4 master-oscillator power-amplifier" at the very beginning of the first sentence. This way of presenting information is somewhat better than the following two ways: "In this paper, we present an efficient…" or "Based on…, this paper presents an efficient…" (Table 3.6)

Table 3.6 Linguistic features of Abstract 3

Sentences	Sections	Tenses	Voices	Styles
S1	Introduction	simple present	passive	impersonal
S2	Methods	simple past	passive	impersonal
S3	Methods	simple past	passive	impersonal
S4	Methods	simple past	passive	impersonal
S5	Results	simple past	passive	impersonal
S6	Discussion/Conclusions	simple present	active	impersonal

(4) The structured IMRD/C abstract (Table 3.7).

Table 3.7 Component analyses of Abstract 4

Sections	Sentences	Original Sentences	Component Analyses
Purpose	S1	The purpose of this paper is to investigate the effect of SiC addition up to 60 percent SiC on the mechanical properties and thermal expansion of Al-7 Vol.% Si/SiC composites.	This sentence introduces the research purpose by "The purpose of this paper is to investigate…"

Chapter 3 How to Write Abstracts

continued

Sections	Sentences	Original Sentences	Component Analyses
Design/ methodology/ approach	S2	Composite specimens containing 7, 15, 30, 45 and 60 Vol.% SiCp are fabricated by pressureless infiltration technique.	These two sentences describe the Methods by "...are fabricated by...technique," "The obtained...are characterized."
	S3	The obtained metal-matrix composites (MMCs) then are characterized for density, porosity, microhardness, ductility, and coefficients of thermal expansion (CTE).	
Findings	S4	The results show that the composite specimens have very low-porosity volumes.	These four sentences summarize the results by using "The results show...", "Moreover, it is found that...are distributed uniformly...," "...decreases with..." and "However, ...tend to increase...and improve ductility."
	S5	Moreover, it is found that silicon carbide particles are distri-buted uniformly in the matrix.	
	S6	Both porosity and mean linear CTE of the composites decrease with silicon carbide volume fraction.	
	S7	However, higher amount of SiCp reinforcement content tends to increase density, microhardness and improve ductility.	
Originality/ value	S8	The processing employed in this paper would enable realization of electronic packages made out of Al-Si/SiCp MMCs.	This sentence emphasizes the contribution by using "...would enable realization of..."
Cited from Tahat, 2010:21.			

Structurally, this abstract follows the structured IMRD/C model. There are altogether eight sentences but four sentences (Sentences 4-7) emphasize the Results section and one sentence summarizes the originality/value (Table 3.8).

Table 3.8 Linguistic features of Abstract 4

Sentences	Sections	Tenses	Voices	Styles
S1	Introduction	simple present	active	impersonal
S2	Methods	simple present	passive	impersonal
S3	Methods	simple present	passive	impersonal

Sentences	Sections	Tenses	Voices	Styles
S4	Results	simple present	active	impersonal
S5	Results	simple present	passive	impersonal
S6	Results	simple present	active	impersonal
S7	Results	simple present	active	impersonal
S8	Discussion/Conclusions	past future	active	impersonal

Linguistically, apart from the uniform simple present tense and the impersonal style, one noticeable feature of this abstract is that the passive voice is used in the Methods section (Sentences 2-3) and the Results section (Sentence 5) to describe the author's proposed studies as well as to emphasize the objectivity of the research. Actually four specific functions of the passive form compared with "we plus an active verb" can be identified in academic writing: ① "we" indicates the author's procedural choice, while the passive indicates an established or standard procedure; ② "we" is used to describe the author's own work and the passive to describe the work of others, unless that work is not mentioned in contrast to the author's in which case the active is used; ③ the passive is used to describe the author's proposed studies; and ④ the use of the active or the passive is determined by focus due to the length of an element or the need for emphasis (Taron, *et. al.*, 1998).

2) The result-driven abstract

The previous four examples have already illustrated the importance of the Methods and Results section in RA abstracts. They deserve more attention as Abstract 1 and Abstract 3 both emphasize the research methods, while Abstract 4 stresses the research findings. However, some journals especially prestigious journals like *Nature* tend to emphasize the Results and the Conclusions in the abstract. Ayers (2008) analyzed RA abstracts in the scientific journal *Nature* and found the Methods have been completely removed and the Results incorporated into the Conclusion became the only obligatory move since 1997 (Table 3.9).

Table 3.9 Component analyses of Abstract 5

S	Original Sentences	Component Analyses
S1	We find an analytical condition characterising when the probability that a Lévy Process leaves a symmetric interval upwards goes to one as the size of the interval is shrunk to zero.	This sentence introduces the result and the conclusion by using "We find..."
S2	We show that this is also equivalent to the probability that the process is positive at time t going to one as t goes to zero and prove some related sequential results.	This sentence describes the result and the conclusion by using "We show... and prove..."

S	Original Sentences	Component Analyses
S3	For each $\alpha > 0$, we find an analytical condition equivalent to $X_{T_r} T_r^{-1/\alpha} \xrightarrow{p} \infty$ and $X_t t^{-1/\alpha} \xrightarrow{p} \infty$ as $r, t \to 0$ where X is a Lévy Process and T_r the time it first leaves an interval of radius r.	This sentence further describes the result and the conclusion by using "...we find..."
Cited from Probab, 2008:103.		

Structurally, this abstract is quite different from the previous four examples as these three sentences are all about the research results and conclusions for they begin with "We find...," "We show that... and prove..." and "...we find..." respectively (Table 3.10).

Table 3.10 Linguistic features of Abstract 5

Sentences	Sections	Tenses	Voices	Styles
S1	Results+Conclusions	simple present	active	personal
S2	Results+Conclusions	simple present	active	personal
S3	Results+Conclusions	simple present	active	personal

Linguistically, the personal style, the active voice and the simple present tense used in this abstract also create parallelism by "we plus verbs" to highlight the authors' contributions. Harwood (2005) claimed that "First person pronouns help writers create a sense of newsworthiness and novelty about their work, showing how they are plugging disciplinary knowledge gaps." (Cited from Livnat, 2012: 93-94)

3.3 Tips on Writing a Good Abstract

Hyland (2004) stated "the marketing of an RA usually begins with the abstract where writers have to gain readers' attention and persuade them to read on by demonstrating that they have both something new and worthwhile to say" (Hyland, 2009:70-71).

Generally speaking, there are four kinds of strategies to write a good abstract, namely strategies at the ***discoursal*** (语篇的) level, strategies at the ***syntactic*** (句子的) level, strategies at the ***lexical*** (词汇的) level and strategies with ***graphics***.

1) **Discoursal strategy**

(1) **Compress or delete unnecessary sections to foreground the results if applicable.**

Most often, for the sake of word limit (50 words) or emphasis of the research contribution, the Background section, the Introduction section, even the Methods section, will be compressed or deleted. Abstract 5 is a good example to illustrate this discoursal strategy as the Background section and the Introduction section have been deleted.

(2) Follow the IMRD/C model or the BPMRC model strictly.

Actually, both the IMRD/C model and the BPMRC model are *rhetorical strategies* (修辞策略) that help to justify the enterprise of experimental science (Gross, 1985; Hyland, 1998). They are both logic structures for persuasion. **The Introduction persuades the reader** that "the research undertaken is necessary and worthwhile on the grounds that there exist some gaps in knowledge on a topic which is important"; **the Method section persuades the reader** that "the research was well done, specifically that the subjects represented the groups they were intended to represent and the experimental method avoided distortion"; **the Results section persuades the reader** that "the statistical packages used were useful and informative"; and **the Discussion persuades the reader** that "the results make sense and fit with other examples of research, leading to a consistent body of knowledge" (Cited from Zohar Livnat, 2012: 27). Actually example 1, example 2, example 3 and example 4 are all constructed in this way whether they are structured abstracts or not.

2) Syntactical strategy

(1) **Use the parallel structure if possible.**

Syntactically, one effective way to foreground the researchers' contribution is to use the parallel structure as shown by Abstract 3 (parallelism created by the passive constructions), and Abstract 1 and Abstract 5 (parallelism created by "we plus verbs").

(2) **Choose the subject strategically.**

As illustrated by the first sentence in Abstract 3, "**An efficient single-frequency Nd:YVO4 master-oscillator power-amplifier** is described," although this sentence is top-heavy, yet it can attract the readers' attention immediately.

3) **Lexical strategy**

In order to highlight the contributions of the research, sometimes the authors strategically use promotional words or phrases in their abstracts, especially in the Results and Discussion/Conclusions sections. For example, in the previous five abstracts, the authors use words and phrases like "...we propose a **new** method...," "our proposed method processes datastreams in an **incremental way** with **small computational cost**" and "... **significantly improves**..." (Abstract 1); "...is **important** for... and is therefore **fundamental** to..." and "produce **viable seed**..." (Abstract 2); "An **efficient**...," "**a simple**... **reasonable agreement with** the experimental work" and "This system **offers a promising**..." (Abstract 3); "The results show... **very low**...," "Moreover,... distributed **uniformly**...," "... **decreases**..." and "**tend to increase**..." (Abstract 4); "...upwards... **is shrunk**..." and "... **is positive**..." (Abstract 5).

4) **Graphical Strategy**

Besides the above three strategies, sometimes graphics in some journals such as *The Journal of Physical Chemistry* can also be used to foreground the contribution. The following example is taken from that journal.

Chapter 3　How to Write Abstracts

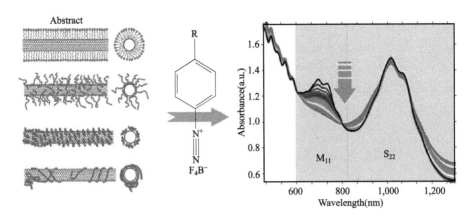

Abstract

<u>Current methods</u> of synthesis for carbon nanotubes (CNTs) <u>usually produce</u> heterogeneous mixtures of different nanotube diameters and thus a mixture of electronic properties. <u>Consequently</u>, <u>many techniques to sort</u> nanotubes according to their electronic type <u>have been devised</u>. <u>One such method involves</u> the chemical reaction of CNTs with aryl diazonium salts. <u>Here we examine</u> the reactions of electric arc produced CNTs (dispersed by a variety of surfactants and polymers in aqueous solution) with 4-bromo-, 4-nitro-, and 4-carboxybenzenediazonium tetrafluoroborate salts <u>in order to find conditions for maximum</u> selectivity. Reactions <u>were monitored</u> through the semiconducting S22 and metallic M11 transitions in the UV_vis_NIR absorbance spectra of the nanotube dispersions. Selectivity <u>was observed</u> to depend heavily on the type of surfactant, the type of diazonium salt and its concentration, the reaction temperature, and the solution pH. <u>Additionally</u>, the surfactant concentration <u>was found to exert a significant influence</u> as the dediazoniation product yields are affected by this parameter. For certain combinations of surfactant and diazoniumsalt the selectivity <u>is markedly improved</u>, particularly in dispersions of nonionic surfactants Pluronic F-127 and Brij S-100, which are similar in structure. Smaller diameter HiPCO nanotubes <u>were better functionalized</u> in dispersions of Triton X-405. The greater selectivity afforded by these poly (ethylene oxide) containing polymers <u>is postulated to</u> arise from electron donation provided by their ether Oxygens. The ionic surfactant sodium dodecyl sulfate <u>was found to display unique behavior</u> in that semiconducting nanotubes <u>were preferentially functionalized</u> at natural pH, likely due charge localization interactions with the surfactant.

(Cited from Blanch, Lenehan, and Quinton; 2012: 1709)

As illustrated by the underlined words and phrases, even without the graphics, this abstract is well written for it employs the corresponding discoursal strategy, the syntactical strategy and the lexical strategy mentioned above.

Firstly it follows the BPMRC model discoursally and presents logically all the necessary information.

Secondly, most of the sentences, especially the Methods section, are written in the parallel structure by the past passive form with the impersonal subject to emphasize the objective and scientific research.

Finally, the results and the implications are highlighted by adjectives, adverbs and verbs like

"... was found to exert a **significant** influence...," "... **is markedly improved**...," "... were **better functionalized**...," "... was found to display **unique** behavior... were **preferentially functionalized**..."

But with the inclusion of the graphics in the abstract, the contribution of the research will be further highlighted as graphics can simplify ideas, can reinforce ideas, can create interests and are universal (Pfeiffer William, 2006:438).

3.4 How to Build Your Own Syntactic Templates

No matter whether you are going to write an/a (un)structured abstract with the IMRD/C or BPMRC model or a result-driven abstract, the most important factor is to know how to construct sentences. For beginners, one effective way to build a database of popular phrases and sentence patterns in your major is to download a certain number of abstracts, delete the technical terms/noun phrases and write down what is left as the core verbs, phrases and sentence patterns. Professor Strozzi, author of *"How to Write a Technical Paper in English—A Repertoire of Useful Expressions"* once stated:

I learned to write engineering papers in English by collecting useful phrases from every article in English that I read. I have used these phrases throughout my career in academia, and I believe that they have served both me and my students very well.

(Cited from Wallwork, 2011: 272)

In the following, some sentence patterns are collected based on the examples analyzed in this chapter and some other handy examples. However, as it focuses on a limited number of samples in different majors, it will not be exhaustive and you are supposed to have your own database following this method.

Syntactic templates in writing the abstract section:
1) Background
Most... research assumes that.... However,... research suggests that...
Previous studies from our laboratory have shown that...
In recent years there has been an increasing interest in... in order to improve...
... has long been used in...
One question/issue that needs to be raised is...

2) Introduction
This paper presents/proposes/ explores /describes/analyzes...
In this paper, we propose/present/explore/describe an... algorithm that uses... in order to...
A new/an efficient... is presented/described/analyzed/investigated/ examined/introduced/ discussed in order to...

3) Methods
The method used in our study is known as...

The... scheme is used to discover...

The technique we applied is referred to...

The procedure they followed can be briefly described as...

The approach adopted extensively is called...

Detailed information has been acquired by the authors using...

4) Results

The experimental results show that... approach yields..., improving up to... over the existing algorithms for...

Results have indicated a significant increase in the performance when compared with...

We carried out several studies which have demonstrated that...

We evaluate...

5) Discussions/Conclusions

Moreover, we show that the robustness of our methods... can be enhanced effectively.

The second contribution of the paper is to introduce a new experiment methodology for...

Overall,... could grant wide applicability in...

... is important for... and is therefore fundamental to...

The results of our experiments demonstrate the potential merits of...

3.5 Summary

In this chapter, we have firstly explored the basic components of RA abstracts by analyzing corresponding examples from five journals sentence by sentence. RA abstracts have been categorized into the summary type and the result-driven type according to the requirements of different journals. In the summary type, in terms of the contents, two subtypes have been identified depending on whether there is a background section in the abstract, namely the IMRD/C model and the BPMRC model. However, in terms of structure, the structured and the unstructured abstracts can be identified. What's more, the linguistic features such as the present tense, the past tense, the active voice, the passive voice, the personal and the impersonal style have been differentiated and their corresponding functions have been analyzed. Next, we have provided tips on writing good RA abstracts from the discoursal level, the syntactic level, and the lexical level. Finally, we have provided the syntactic templates for writing the corresponding sections of an RA abstract—the Background, the Introduction, the Methods, the Results and the Discussions/Conclusions, hoping to provide insights for readers to enhance their academic writing.

Tasks

Task 1

Directions: Read two Abstracts in your research field and identify their basic components as well as linguistic features.

Task 2

Directions: Translate one of the two Abstracts into Chinese and then write down the useful expressions including its syntactic templates.

Task 3

Directions: Write the Abstract section according to the Instructions of your target journal.

Chapter 4
How to Write Introduction

A scientific research paper is often composed of four sections such as introduction, methods, results and discussion (often shortened as IMRaD). Some researchers also include abstracts as one of the five sections.

We preferred IMRaD, in which Introduction section is regarded as the first section of scientific writing and as the most difficult one to be written. The main purpose of the Introduction section is to provide the rationale for the paper, moving from general discussion of the topic to the particular question being investigated. Another purpose is to attract readers' interest in the topic. (Swales and Feak, 2010:156)

As journals differ considerably in terms of their content writing and *conventions* (规范), before planning to write a research article you are required to read the "*Instructions for Authors*" (投稿须知) to identify the basic components in the target journal. Meanwhile, you should read the papers in your field carefully so as to familiarize yourself with the basic components and to notice how the information is expressed and organized.

4.1 Basic Components

In terms of basic components in writing the Introduction section, the requirements of journals vary widely from discipline to discipline and even within the same discipline, basic components might be different. That is to say, some journals might have such components as the research background, methods, results and significance whereas other journals might differ. For example, in the Introduction section of the journal *Optical Engineering*, it first defines key terms and then states the methods.

Different researchers might employ either models, or moves or rules to refer to the structure of the Introduction section. Namely, Glasman-Deal (2011) proposed a model with four basic components: ① Establish the importance of your field. In this component, you as a writer need to provide background information, define the key terms and present the problem or current research focus. ② Describe previous and/or current research and contributions. ③ Locate a gap in the research. Here you need to point out the limitations of other researches and describe the problem you will address. Then you need to present a prediction to be tested. ④ Describe the present paper.

Similarly, Day and Gastel (2007:57) suggested five rules to write a good introduction:

(1) Present with all possible clarity, the nature and scope of the problem investigated;

(2) Review briefly the *pertinent literature* (相关文献) to orient the reader;

(3) State the method of the investigation, reasons for choosing a particular method;

(4) State the principal results of the investigation/the research questions;

(5) State the principal conclusions suggested by the results/show the writing arrangement.

The last component (writing arrangement) often occurs in the long article such as *review* (综述) and some journals of IEEE, which seldom appears in *Letters of Physics* (《物理快报》) or *Science Express* (《科学快报》).

Regarding the length, a big difference also occurs among journals. Some journals might have a very brief introduction like *Letters of Applied Physics* while others might involve a detailed introduction. The next section will provide you with more detailed information via specific analyses.

4.2 Specific Analyses

This section will present you with three examples taken from three different journals, in which specific analyses will be demonstrated with the aim of enhancing your awareness of basic components in writing the Introduction and strengthening your skills of cohesion in macro as well as micro organization.

Please take a look at the first Introduction example taken from *Science Express*, quite brief with 138 words, but this paragraph covers four basic components. For detailed information, please read Table 4.1.

1) Introduction section of E.g. 1 analyses

Table 4.1 Component analyses of the Introduction section of E.g. 1

S	Original Sentences	Component Analyses
S1	Biological dependence on the six major nutrient elements *Carbon* (碳), *Hydrogen* (氢), *Nitrogen* (氮), Oxygen, *Sulfur* (硫), and *Phosphorus* (磷) is complemented by a selected array of other elements, usually *metal* (*loid*) *s* (非金素) present in trace quantities that serve critical cellular functions, such as enzyme co-factors (1).	This sentence presents the scope of the research field by stating biological dependence on six major nutrient elements.
S2	There are many cases of these *trace elements* (微量元素) substituting for one another.	A general description of the previous researches has been made by "...many cases..."

continued

S	Original Sentences	Component Analyses
S3	A few examples include the substitution of *Tungsten* (钨) for *Molybdenum* (钼) and *Cadmium* (镉) for *Zinc* (锌) in some *Enzyme* (酶) families (2, 3) and Copper for iron as an Oxygen-carrier in some *arthropods* (节足动物) and *mollusks* (软体动物) (4).	More specific previous research findings have been revealed by "A few examples..." with the employment of a citation order system.
S4	In these examples and others, the trace elements that interchange share chemical similarities that facilitate the *swap* (交换).	The contribution of the trace elements has been described by "...facilitate..."
S5	However, there are no prior reports of substitutions for any of the six major elements essential for life.	The previous research limitation and the significance of the current research have been stated by "however,... no prior reports..."
S6	Here we present evidence that *Arsenic* (砷) can substitute for *Phosphorus* (磷) in the biomolecules of a naturally occurring bacterium.	The purpose of the current research has been displayed by "... present,... can substitute..."
Cited from Wolfe-Simon, *et. al.*, 2010: 1126.		

This Introduction section is a typical example of a report published in *Science Express*. It focuses on the research background with a general-specific pattern, in which the first sentence just presents the general research background and the next two sentences provide more specific descriptions of the previous researches. The fourth sentence describes the contribution. The fifth sentence displays the research gap, also called limitations, suggesting the significance of the present work. The last sentence has revealed the purpose of the current research.

This Introduction section is combined with six short and long sentences, which have covered the information of four major components as shown in Table 4.1. In spite of many professional words, this Introduction section is very brief, in which you are required to notice how the four basic components are organized. Meanwhile, please identify the various word forms of "substitute" and their usages in different sentences. Another feature is cohesion, which will be analyzed in **Language focus.**

The number(1) has suggested that the idea could be obtained from the *first entry of references* (参考文献第一条). The numbers in *parentheses* (括号) indicate that this paper has employed a citation order system.

(1) Language focus.

E. g. 1 is clearly organized by such linking devices as repetition of key words, conjunction and reference, out of which references and repetition of key words are more popular.

(2) Linking devices.
① Repetition of key words.

The key word "elements" is repeated five times in the four sentences and the word "substitute" with its *derivation*（派生词）"substitution" is also repeated four times. "substitution" in S3 reminds us of "substituting" in the previous sentence. "Substitution" is the repetition of "substituting," which is one type of lexical *reiteration*（复现）. This repetition has efficiently connected the ideas within the paragraph and brought forward the cohesion.

② Conjunctions.

In the initial position of S5, "however" is an **adversative** conjunction, indicating the change of the main idea in the fifth sentence.

③ References.

In S2, "these trace elements" is a **demonstrative reference** which refers back to "the major nutrient elements." The demonstrative reference "these" is an effective cohesive device in organizing text. In S4, there is also a **demonstrative reference** "these examples," corresponding to "a few examples" in S3. "Here" in S6 is also a **demonstrative reference** like "these." It often refers to an extended text, and often has a meaning of "in this respect." "we" in S6 refers to the authors of this paper, and can be considered as a **personal reference**. In a word, every sentence in this paragraph is cohesively combined by different cohesive devices.

2) Introduction section of E. g. 2 analyses

Another example is a common introduction with three or four paragraphs. Now let us take a look at E. g. 2, cited from IEEE 2009, which is presented in three Tables due to three paragraphs (Table 4.2-Table 4.4).

Table 4.2　The first paragraph of E. g. 2 analyses

S	Original Sentences	Component Analyses
S1	Recently, robots driven by air pressure have been widely used in industrial fields because of their low cost, light weight and simple systems [1].	The research background is revealed by stating the research field and literature review.
S2	Also, by using large compliance by air pressure, it is possible to build a safe and flexible drive [2].	The research background is presented with its previous method "by using..."
S3	Many experiments have been conducted recently for supporting human life and for practical applications, such as handling glass.	The research background is described with its purposes by "for... and for..."

The above three sentences have been employed to demonstrate the background information: the research field (robots in industrial fields), the previous method and the previous purposes.

The numbers [1] and [2] indicate that this paper has adopted a citation order system.

Chapter 4 How to Write Introduction

Table 4.3 The second paragraph of E. g. 2 analyses

S	Original Sentences	Component Analyses
S1	To create such robots, it is necessary to develop a humanlike robot hand.	Research objective has been demonstrated.
S2	So, in this research, we use a pneumatic actuator to accomplish this.	Research purpose has been integrated into a method.

The second paragraph is relatively short with two sentences, which have employed three infinitives. The first infinitive "to create such robots" has clearly connected the second paragraph with the first paragraph whereas the other two infinitives have showed the research purpose.

Table 4.4 The third paragraph of E. g. 2 analyses

S	Original Sentences	Component Analyses
S1	There are many robot hands that have been developed using pneumatic actuators [3] [4].	"many robot hands..." has revealed the previous contribution whereas "using..." has indicated the previous method.
S2	However, in recent pneumatic actuators, high air pressure is necessary to have enough output, so a large compressor and system are needed.	The previous research limitation has been stated by using such words as "however, necessary, and needed."
S3	Also, the actuator itself is too large, which is difficult to arrange directly in a human-sized robot hand.	Another research limitation has been described by using such words as "also, too, and difficult."
S4	Consequently, the purpose of this research is to develop a pneumatic actuator with low-pressure and low-volume for driving a robot hand that works flexibly and is safer, on the assumption that it will be in contact with people.	The current research purpose has been introduced by "the purpose..."
S5	We examine the characteristics of this pneumatic actuator, and develop a five-fingered robot hand with pneumatic actuators.	The first research contribution has been presented by "develop..." while the method has been revealed by "with..."
S6	Next, we make a 1-link arm, with one degree of freedom using the pneumatic actuators, and examine the control performance of the pneumatic actuator by constructing a control system for a joint model of the five-fingered robot hand.	The second contribution has been shown by "make..." whereas the method has been demonstrated by "using..., and by constructing..."

In the third paragraph, the first sentence reveals the previous contribution regarding the robot hand whereas the next two sentences (S2 and S3) present the previous research limitations, leading to the purpose of the current research. Then the last two sentences reveal two current research results briefly and integrate the research methods into the results.

This introduction is organized clearly with the combination of short and long sentences. Language used is simple and clear enough for all people to understand, in which the most common structures are *infinitive phrases* (不定式短语) and *prepositional phrases* (介词短语) such as "with...", "by..." and "for..." Please notice that infinitive phrases and prepositional phrases can contribute to the brevity and clarity in a sentence.

Linking devices.

The second introduction section employs more conjunctions as linking devices, which is very popular among Asian writers.

① **Conjunctions.**

In S2 of Table 4.2, "also" is an **additive conjunction**, showing the additive relation between the first two sentences. The initial "so" in S2 of Table 4.3 is a **causal conjunction**. The initial "however" in S2 of Table 4.4 is an **adversative conjunction**, suggesting that the meaning is out of expectation. "Also" in the initial position of S3 of Table 4.4 is a conjunction as well, demonstrating the additive relation between S2 of Table 4.4 and S3 of Table 4.4. S4 of Table 4.4 begins with "consequently," a **causal conjunction**, which means that it is a result of the previous text. "Next" in S6 of Table 4.4 is a **temporal conjunction**, revealing the sequence of two sentences. The temporal conjunction is useful in textual cohesion as well as textual organization.

② **Lexical repetition.**

"air pressure" in S2 of Table 4.2 is a lexical repetition of that in S1 of Table 4.2. S3 of Table 4.2 is linked closely by repeating key information "experiments of robots driven by air pressure." The second paragraph is associated with the first paragraph by repeating the key information "such robots" and "humanlike robot hand" in S1 of Table 4.3 and "to accomplish this" in S2 of Table 4.3. "pneumatic actuators" in S1 of Table 4.4 is a lexical repetition, responding to the same lexical items in S2 of Table 4.3. The third paragraph is also connected with the second paragraph by repeating "robot hands" and "pneumatic actuators" in the six sentences (S1-S6 of Table 4.4).

③ **References.**

"Such robots" in S1 of Table 4.3 is an **anaphoric reference**, referring back to the robots in the first paragraph. And "this" in the last position of the same sentence is a **demonstrative reference**, referring to "a humanlike robot hand" in the previous sentence. "We" in S5 of Table 4.4 refers to the authors of this paper, and can be considered as a **personal reference**. "this pneumatic actuator" in S5 of Table 4.4 is a **demonstrative reference**, reminding us of "a pneumatic actuator" in S4 of Table 4.4.

3) **Introduction section of E.g. 3 analyses**

The last example of the Introduction section is relatively longer, quite common in research articles and review. This Introduction section focuses on the detailed description of materials, which is cited from *Journal of the American Chemistry Society* (Table 4.5).

The first paragraph generally presents us with the research background by introducing the research field: solar cells and two polymeric materials. The last sentence combines the research objective and the method, in which the infinitive "to further improve…" shows the current research objective whereas the verbs "develop, synthesize, or work" reveal the research results (Table 4.5).

Table 4.5 The first paragraph of E. g. 3 analyses

S	Original Sentences	Component Analyses
S1	Solar cells are one key technology for solving world energy needs.	The topic sentence here states the research field.
S2	The development of new materials such as the semiconducting conjugated polymers as active components in bulk hetero junction (BHJ) photovoltaic devices could help to significantly reduce the fabrication cost of such devices.	The background information has been narrowed down to the development of new materials.
S3	For these purposes, two polymeric materials have been extensively studied in the past decade: poly [2-methoxy-5-(3∉, 7∉-dimethyloctloxy)-p-phenylenevinylene] (MDMO-PPV) and regioregularpoly (3-hexylthiophene) (P3HT).	More specific background information has been displayed.
S4	The utilization of these polymeric materials has led to power conversion efficiencies between 3.0% and 5.0% when mixed with [6,6]-Phenyl C61 butyric acid methyl ester (PCBM), as the electron acceptor.	Contribution of the two polymeric materials has been presented.
S5	To further improve the device performances, one can develop new device architectures, synthesize new polymers and new electron acceptors, or work on both ends.	The research objective and the results have been stated.

The second paragraph also presents us with the research background by briefly introducing the previous development of polymer structures, and internal charge transfer (ICT) with specific data. The specific descriptions of ICT can help readers obtain the information that ICT has its advantages and disadvantages in developing new polymers, suggesting the research gap (Table 4.6).

Table 4.6 The second paragraph of E. g. 3 analyses

S	Original Sentences	Component Analyses
S1	In the past few years, several groups of chemists proposed new polymer structures as alternatives to P3HT and MDMOPPV since the performances of these two polymers are somehow limited by their relatively large band gap.	This topic sentence introduces new polymer structures proposed by previous researchers.

continued

S	Original Sentences	Component Analyses
S2	For many years, internal charge transfer (ICT) from an electron-rich unit to an electron deficient moiety has been extensively used to obtain low band gap conjugated polymers.	Contribution of ICT has been stated in developing materials.
S3	Using an ICT strategy, new polymers have been developed to better harvest the solar spectrum, especially in the 1.4–1.9 eV region.	More specific contribution of ICT has been described.
S4	Several low band gap polythiophene derivatives have been reported by Krebs and Reynolds, but until now, relatively low performances in solar cells have been obtained.	The limitation of polythiophene derivatives has been introduced.
S5	Polyfluorene derivatives show promising features with power conversion efficiencies (PCE) between 2.0% and 4.2%.	Potential functions of polyfluorene derivatives have been stated.
S6	However, these polymers show relatively low carrier mobility, limiting the device performances.	The limitation of the polymers has been presented.
S7	Recently, benzothiadiazole and cyclopentadithiophene copolymers showed very interesting PCE values (3.2%–5.5%) and high carrier mobility ($10^{-2}-10^{-1}$ cm$^2 \cdot$ V$^{-1} \cdot$ s^{-1}) demonstrating that one can synthesize ICT polymers having low band gap and high carrier mobility.	ICT polymers' values and contributions have been described.

The third paragraph mainly focuses on the description of materials by briefly stating three features of carbazole derivatives and their limitations for BHJ solar cells. As a result, a new polycarbazole derivative (PCDTBT) is proposed by the current authors (Table 4.7).

Table 4.7 The third paragraph of E.g.3 analysis

S	Original Sentences	Component Analyses
S1	Poly(2, 7-carbazole) derivatives are other excellent potential candidates for BHJ solar cells.	This topic sentence introduces materials for BHJ solar cells.
S2	Indeed, as observed with poly(2, 7-fluorene)s, the physical properties of poly(2, 7-carbazole)s can be easily modulated.	The materials' first feature is presented.

continued

S	Original Sentences	Component Analyses
S3	As an electron-rich molecule, the carbazole unit is perfect for the development of ICT polymers.	The materials' second feature is shown.
S4	Moreover, poly(N-vinylcarbazole) (PVK) is among the best photoconductive polymeric materials.	The materials' third feature is introduced.
S5	Initial studies by K. Mullen and our group have produced polycarbazole materials that exhibit relatively low efficiency in solar cells (0.6%–0.8%).	The limitation of the polycarbazole materials is stated in solar cells.
S6	In general, those polymers were poorly soluble and showed a lack of organization.	The limitations of the previous materials are described.
S7	However, we recently reported a new polycarbazole derivative (PCDTBT) bearing a secondary alkyl side chain on the nitrogen atom of the carbazole unit that shows high solubility and some organization, resulting in a very good PCE (3.6%).	A new polycarbazole derivative (PCDTBT) is proposed.
S8	It was concluded that, by improving the electronic properties of the polymeric materials, much higher efficiencies could be reached.	A new way of developing polymeric materials has been revealed.

The fourth paragraph has focused on the method description: models and specific data. The first sentence just introduces general information of the research method: models and the research's theoretical basis. Then specific information is presented with six sentences by describing three conditions and their ranges in satisfying the requirements, which gradually directs the readers to the research objective in the last sentence(Table 4.8).

Table 4.8 The fourth paragraph of E. g. 3 analyses

S	Original Sentences	Component Analyses
S1	Recently, several models have been proposed to estimate the polymer performance in BHJ solar cells.	This topic sentence briefly introduces methods, the theoretic basis (理论依据).
S2	First, the polymer must be air-stable; therefore, the HOMO energy level needs to be below the air oxidation threshold (ca. −5.27 eV or 0.57 V vs SCE).	The requirement of developing the polymer has been stated via a specific description of the HOMO energy level.
S3	Furthermore, this relatively low value assures a relatively high open circuit potential (V_{OC}) in the final device.	One reason has been given in satisfying the requirement of developing the polymer.

continued

S	Original Sentences	Component Analyses
S4	Second, the LUMO energy level of the polymer must be positioned above the LUMO energy level of the acceptor (i.e., [6,6]-Phenyl C61 butyric acid methyl ester (PCBM)) by at least 0.2–0.3 eV to ensure efficient electron transfer from the polymer to the acceptor.	Another condition has been introduced specifically.
S5	Therefore, the ideal donor LUMO energy level should be between −3.7 eV and −4.0 eV.	The range has been described at the LUMO energy level in the second condition.
S6	Finally, the optimal band gap, considering the solar emission spectrum and the open circuit potential of the resulting solar cell, should range between 1.2 eV and 1.9 eV.	The third condition has been given in presenting the range of the band gap.
S7	Therefore, the ideal polymer HOMO energy level should range between −5.2 eV and −5.8 eV.	The range has been described at the HOMO energy level in the third condition.
S8	Taking into account the LUMO energy level of PCDTBT (−3.60 eV), one needs to reduce this parameter to further increase the performance of a polymeric cell while keeping the HOMO energy level within the same energy value (i.e., −5.45 eV).	The research objective has been stated by "to further increase the performance..."

Please note how to describe the conditions and the ranges in satisfying the requirements while reading this paragraph. Try to use them in your future's writing.

The fifth paragraph has employed four sentences to demonstrate how the current research has been conducted, in which previous theoretical models and data were referred to so as to make quantum calculations, compare and test the values of the polymers in solar cell devices (Table 4.9).

Table 4.9 The fifth paragraph of E.g.3 analyses

S	Original Sentences	Component Analyses
S1	On the basis of these initial data and theoretical models, we investigated several alternating polymeric structures to develop the most suitable poly(2,7-carbazole) derivatives for BHJ solar cell devices.	The method was revealed by "on the basis of..." and "several alternating polymeric structures" whereas "to develop..." is the result of the research.
S2	To estimate the polycarbazole performances using those models, we performed quantum calculations on the repeat unit of the planned polymers to determine both the HOMO and the LUMO energy levels.	The current research method was introduced by "using those models, we performed..."

continued

S	Original Sentences	Component Analyses
S3	These values were then compared to the experimental values of the polymers.	The current research method was stated by "compared."
S4	These new polymers were tested in solar cell devices, and their performances were analyzed in terms of polymer organization, molecular weight, charge carrier mobility, and LUMO energy level.	The current method was revealed by "tested…,… analyzed…"
Cited from Blouin, *et. al.*, 2008: 732		

To sum up the third Introduction section, we can conclude that it is a popular one structured from broad information to the specific (narrower and narrower) information with five paragraphs. The first paragraph generally introduces the research field solar cells (a big picture), next states the development of new materials (less general), then describes more specific materials such as two polymeric materials and new device architectures. The second paragraph states the recent materials development in this research field. Then the third paragraph narrows down to poly derivatives as other excellent potential candidates for BHJ solar cells in spite of its limitations. The fourth paragraph presents the key points of the research: theoretical models with data of the previous research and the objectives of this research. The fifth paragraph provides the methods and results.

This Introduction section is also a good example of clear organization by using the topic sentence at the beginning of each paragraph, in which ideas are well-developed by providing specific evidence. Besides, short and long sentences are combined and the words are simple and clear.

Other journals with similar lengths are *Automatica*, *Neurocomputing*, *Transactions on Automatic Control*, *Transactions on Industrial Informatics* and *Journal of Guidance, Control and Dynamics*. Some journals might have a much longer Introduction section. For example, AIAA Journal.

Linking devices.

This Introduction section is characterized by three popular linking devices, which will be described in details as follows.

① Conjunctions.

Two **adversative conjunctions** (but, however) have been employed in S4 and S6 of Table 4.6 of the second paragraph. As an **additive conjunction**, "moreover" in S4 of Table 4.7 shows the additive relation between S3 and S4 of Table 4.7. There are two cohesive devices used in S7 of Table 4.7: The initial "however" is an **adversative conjunction**, showing the meaning is contrary to the expectation; "we" is a **personal reference**, referring to the authors of this paper. "First" in S2 of Table 4.8 is a **temporal conjunction**, corresponding to "several models" in the previous sentence, which shows that this is the first model. Such device is also used in S4 and S6 of Table 4.8 ("second" and "finally," respectively). These temporal conjunctions demonstrate the sequence of the three sentences. As an **additive conjunction**, "furthermore" in S3 of Table 4.8 reveals the additive relation between S2 and S3 of Table 4.8. "Therefore" in S5 and S7 of Table 4.8 is a **causal conjunction**, suggesting the casual relation between sentences.

② **Lexical repetition.**

Key words can be identified frequently especially in the first two paragraphs of this Introduction section, coherently constructing a cohesive chain across sentences.

③ **References.**

In S3 of Table 4.5, the **demonstrative reference** "these purposes" refers back to the meanings in the previous two sentences. The demonstrative phrase "these polymetric materials" in S4 of Table 4.5 is a **demonstrative reference**, which generalizes the materials in the previous sentence. "other" in S1 of Table 4.7 is a **comparative reference**, showing the difference compared with the previous sentences. "Those polymers" in S6 of Table 4.7 is a **demonstrative reference**, referring back to S5 of Table 4.7. In S8 of Table 4.7, the item "higher" is a **comparative reference**, combining the sentence with the previous ones. "these initial data and theoretical models" is a **demonstrative reference** in S1 of Table 4.9, which can be retrospected from S2 to S8 of Table 4.8. "those models" in S2 of Table 4.9 is another **demonstrative reference**, referring back to the previous sentences. Moreover, "these values", a **demonstrative reference** in S3 of Table 4.9, can be retrospected in S2 of Table 4.9. "these new polymers" in the last sentence refers back to the polymers in S3 of Table 4.9, which is also a **demonstrative reference**.

4.3 Tips on Writing a Good Introduction

As we mentioned at the beginning of this Chapter, Introduction section is the most difficult one in writing an academic paper. In spite of its difficulty, you can improve your writing with the help of the following four tips.

1) **How to read papers in your target journal**

As soon as you have a plan to publish your research work, you should read the *Instructions to Authors* of your target journal first to identify basic requirements. Then you need to read highly-cited papers in your research field to identify what other researchers have done and what others have not done. That is to say, you can describe the research background successfully in the Introduction section only when you know quite well what contributions and limitations have been made by other researchers in this research field. In addition, you need to familiarize yourself with basic components in the process of reading the Introduction sections.

2) **How to write accurately, briefly and clearly**

Obtaining enough knowledge in the research field is not enough. You have to learn how to put your information accurately, briefly and clearly. The best way is to imitate, which is also recommended by Day and Gastel (2007: 12). Effective imitation requires noticing the macro structure, including basic components and their lengths, and the micro structure like linking devices, and idiomatic expressions such as professional words, and syntactic templates when you read highly regarded scientific papers in your research field. Another effective imitation can be obtained either by translating a paper written by native writers, or by writing down useful expressions in a notebook. But when you imitate, you should avoid copying the entire sentences.

3) When to write

To write the Introduction, you need to make sure of what you have done and what you have found in your research. So the best time to write it may be after you describe your Methods section or even after you finish your Results section.

4) How to impress the readers and reviewers

The Introduction section plays an important role in getting your research work published. Thus you need to provide clear and brief background information, enough relevant citations, methods, research gaps and distinguished significance of your research work. Moreover, you need to organize all the necessary information briefly and clearly.

4.4　Common Syntactic Templates

We all know the importance of imitation in academic writing, which can ensure to write idiomatically. The following syntactic templates and expressions are just a few examples cited from some Introduction sections *for your reference* （供你参考）, which can be far from being *comprehensive* （包罗万象的）. Thus, you can write down popular syntactic templates and useful expressions in your notebook while you are reading literature in your research field.

1) Introduction of the research background

This paper considers...;

Recent research efforts have been directed toward...;

...has ignited wide-spread interest because of its unique ability to...;

In response to this concern, recent research efforts have been devoted to the development of...

2) Reasons for conducting the research

Despite the many advantages conferred by..., there also remain several drawbacks to its usage.

The most obvious disadvantage of... is that...

One common issue with... is that...

Previous work has been unable to address the state constrained problem.

With our new proofs, more difficult problems can be solved and earlier results can be obtained as special cases.

The earlier results only handle cases where the optimal state trajectory touches the state constraint boundary at most a finite number of times.

... address the following questions, find answers to the following questions; bridge the gap between... and....

3) Literature review

A few examples include the substitution of tungsten for molybdenum and cadmium for zinc in some enzyme families (1, 2) and copper for iron as an Oxygen-carrier in some arthropods and mollusks (3). (Citation Order System)

Several works (Marden, Arslan, and Shamma, 2009; Song, Ding, Kamal, Farrel, and Roy-Chowdhury, 2011; Zhu and Martínez, 2013) have also considered situations where individual agents have objective functions of their own that are aligned with a global objective. (Name and Year System)

4) Conclusion (significance of the research)

Our approach also differs from mainstream MPC by computing all control gains offline. In contrast, MPC requires the (repeated) solving of an optimization problem online; our controller requires only selecting the appropriate controller gain based on the observed switching sequence.

Our framework for... can be used as the basis for...;

The results of this paper can be applied to...;

The remaining contributions of the paper concern...;

The main contributions of the present paper are...;

These novel systems represent combinations of...;

It is often desirable to use...;

...concept is centered on...;... provide insight for...;

...have significant impacts on...;... has emerged into a large market...;

...has brought about the growth of these research efforts.

5) Organization for the paper

The rest of this paper is organized as follows: Section II gives a brief introduction to the modeling of fingerprint minutiae in terms of statistical mixture models. Section III discusses the log-linear random effects model that is used to study how the PRCs change as a function of the underlying image quality. Section IV presents the methodology for fitting the log-linear models in a Bayesian framework. Section V presents the fitting of the models to two databases, namely, the IBM ground truth and NIST fingerprint databases. Section VI presents the summary and conclusion. (Dass, 2010:63)

...Section 2 formulates the optimal leader allocation problem. Section 3 unveils some key properties of the cost—to target functions and optimal leader allocations. Sections 4 and 5 deal with the no-cost and costly switching cases, respectively. Section 6 gathers our conclusions and ideas for future work. (Richert and Cortés, 2013)

Other useful expressions might be: structure, report, demonstrate, reveal, *define* (定义), develop, *generalize* (概括), and outline.

4.5 Summary

We presented major components of an Introduction section and analyzed three examples from three different journals, with the aim of enhancing the readers' awareness of basic components in writing the Introduction section and helping them improve their writing skills by analyzing the linking devices in the three examples and providing four tips as well as syntactic templates and useful expressions cited from Introduction sections of various journals. In terms of linking devices, lexical repetition and references are more frequently employed in scientific paper writing.

Tasks

Task 1
Directions: Read two Introduction sections in your research field and identify their basic components as well as linking devices.

Task 2
Directions: Choose one of the two Introduction sections to be translated into Chinese and then write down useful expressions including its syntactic templates and linking devices.

Task 3
Directions: Write one Introduction section according to the *Instructions to Authors* of your target journal.

Chapter 5
How to Write Methods

In a scientific research paper of IMRaD, the Methods section is of great concern. In different journals or different disciplines, researchers may employ different terms for Methods section, among which "Materials and Methods," "Methodology," "Procedures," "Experiments," "*Experimental*" (实验部分), "Models," and "*Simulations*" (模拟实验) are frequently seen on the basis of the actual needs of their papers.

This section presents a detailed description of the procedure, the equipment and the facilities in your study, whether it is a design for a survey study, an experimental research, a field investigation, or even a combination of several of these.

The purpose of this section is:

➤ to present a detailed description of the design of an experiment or an investigation;

➤ to provide with careful instructions so that the process of your experiment or investigation can be easily visualized and replicated;

➤ to provide enough information for the readers of your paper to judge whether your experimental methods are appropriate or not.

Therefore, it is rather vital for you to meet the requirements of explicitness, transparency and sufficiency in detail when you write this section. The cornerstone of the scientific research requires that the results of your research must be reproducible, and you must provide the basis for the repetition of the experiment or investigation by others in order for the results to be adjudged reproducible. When your paper is subjected to peer review, a good reviewer will read this section carefully. Once he feels doubtful that your experiment could be repeated, the reviewer will pose serious rejection of your manuscript (Day and Gastel, 2007).

5.1 Basic Components

When writing the Methods section, the author usually follows the procedures in the experimental or investigational sequence, as is shown in Table 5.1.

Table 5.1 Basic components of Methods section

1. Generalization or Introduction;
2. Materials or Subjects;
3. Methods or Procedures;
4. Data Analysis.

Sometimes this writing order may not be necessarily followed, because different requirements may occur in a certain discipline or in a certain journal. If your study involves two or more designs simultaneously, the several designs are always described one after another.

5.2 Specific Analyses

Now, we will discuss how the four components construct the Methods section. The purpose is to familiarize you with the basic structure of Methods section and to help you grasp the skills to compose all the details into this section.

1) Generalization and introduction

This is a general introduction of the experiment or investigation, sometimes also presenting the purpose of the study. Usually it follows some background information or the author's hypothesis to the research.

2) Materials

Materials may refer to Instruments or Facilities, i.e. your tools used to measure dependent and independent variables in your study. Materials may also be subjects, which refer to the animals or plants, even the people, that were studied and measured with your tools or instruments and whose responses were used to answer your research questions.

(1) Materials of an experiment study.

In an experiment study or test, the experimental/testing conditions are encouraged to be given, including a careful description of the testing/experimental system, the detailed instructions for the testing/experimental process, the temperature and humidity of the test/experiment, the scope of the speed or pressure required and so on (Day and Gastel, 2007).

In writing Materials section, the exact technical designations and quantities of the materials, the source of or the way to prepare the materials, and sometimes the pertinent chemical and physical properties of the materials should be included.

Generally speaking, the specifications of a chemical or a mechanical/electronic device and meter are provided, including its generic or chemical name, geometric parameters, functions, operating methods, and the name of its manufacturer. Usually the trade name of a material is not preferred to avoid its advertising inherent in the trade name. If it has to be used, the trade name of a material should be capitalized (Teflon, for instance) to distinguish it from its generic name.

Materials like polymeric compounds, ceramics or synthesized materials should be given the ingredients, structure and function related to the paper, even the manufacturer. If there are too many parameters for the materials, a table will be used.

A simulation then requires the software, the formulae used in the experiment. For some special characterizations, such as the *inhibiting release* (控制释放) of a certain medicine, or the *critical dissolving temperature* (临界溶解温度), the specific definition and the calculating equations should be given (Table 5.2).

Table 5.2 Example of materials and its analyses

S	Original Sentences	Component Analyses
S1	In this letter we present the first systematic study on the electrical and magnetic effects of hole compensation in $Ga_{1-x}Mn_xP$.	This sentence shows Purpose 1 of the experiment.
S2	We utilize the amphoteric nature of native defects[13]—donorlike in $Ga_{1-x}Mn_xP$ (Refs. 6 and 14)—to investigate a very wide range of P without significantly changing x.	This sentence reveals the method to achieve the purpose.
S3	A similar method has recently been applied to $Ga_{1-x}Mn_xAs$,[15] and we find surprising similarities between the materials despite the radically different degrees of hole localization.	This sentence presents a comparison of the current study and a previous research.
S4	Furthermore, we present a picture for hole conduction by variable-range hopping (VRH) in $Ga_{1-x}Mn_xP$.	This sentence shows Purpose 2 of the current study.
S5	The samples for this study were prepared by II-PLM.[16]	S5 is a generalization of the sample preparation.
S6–S8	A GaP (001) wafer-doped n-type; $n - 10^{16} - 10^{17}$ cm^{-3}—was implanted with Mn^+ at an energy of 50 keV and an angle of incidence of 7° to a dose of 2×10^{16} cm^{-2}. Samples with approximate side lengths of 6 mm were cleaved along ⟨110⟩ directions and individually irradiated with a single~0.4 J cm^{-2} KrF laser pulse (248 nm wavelength, 18 ns full width at half maximum), homogenized to a spatial uniformity of ±5% by a crossed-cylindrical lens homogenizer. They were subsequently subject to 24 h HCl etching to remove residual surface damage.	These three sentences introduce the preparation of the samples by employing passive voice.
S9 and S10	These parameters have been used previously to produce samples with $x \approx 0.038$.[8] For our samples, x is defined as the peak substitutional manganese (MnGa) fraction—occurring between 20 and 30 nm below the surface—as determined by a combination of secondary ion mass spectrometry (SIMS) and ion beam analysis (IBA).[17]	These two sentences show the readers the difference in the parameters used in the current study and the previous studies (S5) and justify the current parameters by using "as determined by..."

continued

S	Original Sentences	Component Analyses
S11	Compensating defects <u>were then introduced</u> into samples by consecutive irradiations with Ar+ at an energy of 33 keV and an angle of incidence of 7°, which according to simulation 18 yield a vacancy depth profile similar to the typical Mn distribution.	This sentence indicates the method of the sample defect compensation, describing the condition by using "by consecutive irradiation with..."
Cited from Winkler, *et. al.*, 2011.		

Language Focus:

A. Use of first person.

The use of first person "we" avoids using so many passive-voice sentences, which seems rather arbitrary sometimes, and it is easier for the readers to accept the authors' idea.

B. Use of present tense.

The simple present tense indicates the current research as in most sentences in this example, while the present perfect tense shows a previous study as in S3.

C. Use of the simple past tense.

It is much obvious to see that there is a change in tense use from the simple present tense to the simple past tense to indicate what the authors did in their current study. In this process, the simple present tense frequently indicates some conventional scenario while the present perfect tense suggests something that happened previously.

D. A lot of use of passive voice.

Another obvious change from the previous paragraph of the generalization is the use of passive voice, suggesting that the authors try to describe the process of their study in an objective way.

E. Conditions to prepare the materials.

When it comes to the conditions of preparing the materials, detailed description is needed. The temperature, the pressure, the density, anything required should be clearly listed.

(2) Materials for chemical experiments.

In materials chemistry study, the experiment mainly includes *materials*, *synthesis* (合成方法), and *characterization or measurements* (表征部分). If chemicals or reagents are used in the experiment, their proper information should be given: the generic or chemical names, their chemical and analytic purity, the manufacturers, and the like. Sometimes it is even necessary that the molecular structure of a chemical be provided, and the need of further purification of a reagent be indicated.

The following example is chosen from a paper in *Journal of Materials Chemistry* of RSC (Royal Society of Chemistry) in Britain (Table 5.3).

Table 5.3 Example of materials and its analyses

S	Original Sentences	Component Analyses
S1-S8	γ-cyclodextrin (CD) (**WAKO, Japan**) was used as received without further purification. β-CD (**Sinopharm Chemical Reagent Company, China**) was recrystallized three times before use. N-isopropylacrylamide (NIPAAm) (**Acros, Belgium**) was purified by recrystallization from n-hexane. Poly (propylene glycol) bis (2-aminoproyl ether) (PPGBA) were purchased from **Sigma, USA**. 2-Brompropionyl bromide was available from **Alfa Aesar, USA**. Triethylamine (TEA) (**VAS Chemical Reagents Company, Tianjin, China**) was *refluxed* with p-toluenesulfonylchloride and *distilled* under vacuum. Copper(I) chloride [Cu(I)Cl] was *prepared from* $CuCl_2$, *purified* by stirring in acetic acid, *washed* with methanol and finally *dried* under vacuum prior to use. All other solvents and reagents were of analytical grade.	1. The producers or suppliers of the chemicals and reagents are provided (marked in bold letters) 2. The processing methods are clearly given (marked in italics).
Cited from Mater, 2011.		

Language Focus:
Use of abbreviations and proper nouns.

The abbreviations are usually those names for the chemicals or reagents used in the experiment and the proper nouns are used to indicate the producers or the manufacturers and their countries.

3) Methods or procedures

This is a data-collecting section, presenting the experimental or testing method. A new method requires the information about collecting data in detail and in order. To be precise is the key. If you are not sure whether the reader knows the details of what you did, provide as much information as possible. Try to involve each detail. For instance, if a reaction mixture is heated, give the temperature. While if an object is pressed, give the pressure. Questions of "how" and "how much" should be answered precisely to make your results clear and correct. And your accurate and careful presentation of each step of your experimenting or testing can also indicate that you are a capable, careful and qualified researcher.

If your method has been published in a journal, the literature reference is needed. Or if it is a well-known method to your readers, just mention its name.

Now we will analyze several examples chosen from papers on statistical algorithm study, and chemical experimental study.

(1) Methods in statistical estimation and algorithm study.

Table 5.4 is one example adopted from the *Journal of Guidance, Control and Dynamics*.

Chapter 5 How to Write Methods

Table 5.4 Example of methods and its analysis

S	Original Sentences	Component Analyses
S1	**III. Estimation Methods** Two candidate nonlinear estimators, an EKF and UKF, <u>are implemented</u>.	What will be used to do the estimation?
S2	The two filters use the same steps to obtain their estimates but differ in their calculations of certain variables and probability density function (PDF) assumptions within those steps.	What is the difference between the two filters used for the estimation?
S3	Specific details as to how the UKF and EKF differ in their methods of calculating variables within the sequential Kalman filter algorithm <u>can be found</u> in the works of Julier and Uhlmann [8] and Gelb *et. al.* [17], respectively.	What is the source of the difference between the two filters?
S4	In both cases, a recursive algorithm <u>is used</u> that provides optimal estimates for the state ($\hat{X}_{i,k}^*$) and covariance ($\hat{\underline{P}}_{i,k}^*$) of each object at each simulation time step, <u>given</u> by $$\hat{X}_{i,k}^* = \begin{bmatrix} \hat{x}_{i,k}^* \\ \hat{y}_{i,k}^* \\ \hat{\dot{x}}_{i,k}^* \\ \hat{\dot{y}}_{i,k}^* \end{bmatrix} \quad (8)$$ And $$\underline{P}{-}_{i,k}^* = E[X_{i,k} - \hat{X}_{i,k}^*][X_{i,k} - \hat{X}_{i,k}^*]^T \quad (9)$$ where the <u>estimated</u> variances ($\hat{\sigma}^2$) of each state variable <u>are given</u> along the diagonal elements of the covariance estimate.	How can the optimal estimates be obtained? (1)
S5	Within each simulation time step, both estimators will undergo a prediction step where estimates <u>are propagated</u> forward in time to yield prediction state ($\hat{X}_{i,k+1}^p$) and covariance ($\hat{\underline{P}}_{i,k+1}^p$) estimates.	How can the optimal estimates be gained? (2)
S6	This step <u>is followed</u> by an update step where new optimal estimates <u>are obtained</u> through observation.	What is the next step?
S7	For each filter, <u>given</u> prediction estimates ($\hat{X}_{i,k+1}^p$) and ($\hat{\underline{P}}_{i,k+1}^p$), along with current measurements $y_{(i,j),k+1} \,\forall j \in S_{i,k+1}^\tau$, new optimal estimate updates <u>are computed</u> for all objects <u>tasked</u> to be observed (methods for determining objects <u>tasked</u> for observation in the set O_{k+1}^τ, as well as the sets of sensors <u>tasked</u> to observe them $S_{i,k+1}^\tau$, <u>are detailed</u> in Sec. IV).	How are the new optimal estimate updates obtained?
	Cited from Williams, *et. al.*, 2013.	

Language Focus:

A. Unique tense use.

In this example, we can only read sentences in the simple present tense, unlike those in the experiment and investigation studies with a change of several tenses, because such kind of study focuses on statistical analyses and calculation rather than the description of a process. The unique use of simple present tense presents the truth of the statistical analyses and calculation.

B. Use of functions and algebraic expressions.

Functions, equations and algebraic expressions appear during the statistical analyses and calculation. However, not all the functions, equations and algebraic expressions should be recorded in your paper. You should only record those most vital and critical to the final result of your study.

(2) Methods in chemical experimental study.

As for chemical experimental study, the experiment includes materials, synthesis, and characterizations or measurements. The materials part is discussed in the previous section. Now we'll discuss the Synthesis and the Characterization or Measurements sections.

A. Synthesis.

This section describes the process of the experiment. One thing ought not to be ignored. After describing the method or procedures of the synthesis, you have to give the *NMR value* (核磁数据). The FT-IR analysis or chemical element analysis should be listed and the target product should be given its productivity.

B. Characterizations or Measurements.

This part deals with the methods to characterize the reagent, the name and the model of the device used, the parameters set for the test, and the way to prepare the sample reagent. For more information, please look at Table 5.5.

Table 5.5 Example of characterizations and its analyses

S	Original Sentences	Component Analyses
S1–S10	**Synthesis of a distal 2-bromopropionyl end-capped Pluronic F127 (BrP-F127-PBr)** Pluronic F127 was converted to the corresponding ATRP macro-initiator by the end-capping reaction with a fourfold molar excess of 2-bromopropionyl bromide in CH_2Cl_2 using the similar method as described previously[37]. ^1H NMR analysis was used to calculate the degree of esterification (>98%). ^{1}H-NMR (DMSO-d_6): δ4.23–4.24 [s, 4 H, $-CH_2-O-C(=O)-$], 1.02–1.04 (d, 210 H, $-O-CCH_3-C-O-$), 1.71–1.73 (d, 3 H, CH_3-C-Br) ppm.	The way to note down the value is as follows: first, write down the name of the experiment. Here in this example, the name of the experiment is ^1H NMR (核磁氢谱). Next, put down the chemical dissolvant clearly, that is (DMSO-d_6) here. Third, write down the chemical shift values (δ is used to indicate chemical shift): δ 4.23–4.24 [s, 4 H, $-CH_2-O-C(=O)-$], 1.02–1.04 (d, 210 H, $-O-CCH_3-C-O-$), 1.71–1.73 (d, 3 H, CH_3-C-Br) ppm.

continued

S	Original Sentences	Component Analyses
S1–S10	The wide-angle X-ray diffraction (WXRD) measurements <u>were conducted</u> with powder samples using *Philips X'Pert Pro diffractometer* with an *X'celerator detector in a reflection mode*. The radiation source <u>used was Ni-filtered</u>, *Cu Kα radiation with a wavelength of 0.154 nm*. The voltage <u>was set</u> to be *40 kV and the current 20 mA*. Samples <u>were mounted</u> on a sample holder and <u>scanned</u> from *4.5° to 60° in 2θ at a speed of 5 deg/min*. The high-resolution transmission electron microscopy (TEM) micrographs <u>were taken</u> with a *JEM – 2010 electron microscope operated at 200 kV accelerating voltage*. **The sample <u>was prepared</u> by dropping a few microlitres of the aqueous solution of the PR (0.3 mg/ml) onto the Formvar-carbon film-coated copper grid. Excess solution <u>was quickly wicked</u> away with a filter paper.**	In this example, first the name of the device or instrument should be given (words in italics), then their parameters or experimental conditions follow (words in italics). The boldfaced letters indicate the preparation of the sample.
Cited from Wang, *et. al.*, 2010.		

Language Focus:
Use of abbreviations and chemical expressions.

The abbreviations are usually those names for the chemicals or reagents used in the experiment or the unit of a certain reagent and the chemical expressions are used to indicate the structures of the chemicals.

4) Data-analyses

The Data-analysis section deals with the procedures of analyzing the data collected in the previous section and verifies its validity. Actually, in a scientific research paper, especially a lengthy enough one as in *Journal of Guidance, Control, and Dynamics*, data-analysis can easily be seen all the way along the data-collecting process, because any data are obtained through test-calculation-analysis process and every time when certain data are obtained and analyzed, new parameters or variables appear, which follows a new round of test, calculation, and analysis. Whereas when the final data are achieved, there is always a general analysis of all the data obtained in order to suggest that all the data obtained are valid and the writer's research methods are effective. After certain data are analyzed, there is often an expression indicating or implying that a further problem exists and should be considered or solved later.

Data analysis can be quantitative or qualitative. For a quantitative data-analysis, the statistical method needs to be mentioned. The ordinary statistical methods can be used without any comment, while the advanced or unusual methods do need to be cited. If it is a qualitative data-analysis, the procedures should be specified and categories deriving from the data should be addressed.

Here is an example of data analyses from *Journal of Guidance, Control, and Dynamics* (*Table* 5.6).

Table 5.6 Example of data analysis and its component analyses

S	Original Sentences	Component Analyses
S1	**1. Effects on Performance due to Λ and t_F^*** (S1) The first set of simulations is directed at analyzing the effect of the guidance matrix Λ on the guided trajectories.	The first sentence suggests the purpose of the first set of simulation.
S2 and S3	(S2) For this specific case, t_F and t_F^* are fixed. (S3) The simulations are initiated at time $t=0$ and conducted until a fixed final time selected to be $t_F = 7,778.3$ s.	This is the condition of the simulation.
S4–S18	(S4) Assuming that the guidance matrix is spherical (i.e., $\Lambda = \Lambda \mathrm{diag}\{1,1,1\}$), the trajectory and the acceleration command histories depend on only one single guidance gain Λ as well as n. (S5) Assuming a fixed value of $n = 0.5$ (i.e., the second sliding surface is reached in half of the total flight time), the guidance gain Λ is varied parametrically. (S6) Four MSSG-guided trajectories have been simulated with Λ equal to 1.5, 2, 4, and 8, respectively. (S7) Figure 1 shows the three-dimensional (3-D) guided descent and landing trajectories, as well as the history of the position components along the asteroid-centered body-fixed coordinates. (S8) Figure 2 shows the history of the three velocity components. (S9) Figure 3 shows the history of the three components of the acceleration command. (S10) The MSSG algorithm generates trajectories that can be subdivided in two phases. (S11) In the first phase of the flight ($0 < t < t_F^*$), the acceleration command drives the second sliding surface to zero. (S12) Once the second surface is reached ($t = t_F^*$), the second phase is initiated. (S13) During this phase, the first surface is driven to zero according to the nonlinear first-order dynamics dictated by Eq. (18). (S14) For each of the three components, the magnitude of the acceleration command tends to increase with the parameter Λ, which regulates the rate at which the first surface is reached. (S15) Although the time of flight is fixed i.e., the first surface is reached exactly at ($t = t_F$), the rate of convergence depends on Λ. (S16) Figure 4 shows the time history of the first and second	These sentences show the calculation and reasoning. Typical sentence patterns are used here: 1. To indicate mathematical hypothesis: "Assuming (that)..." (S4 and S5), "Once..." (S12); 2. The present perfect tense—"have been simulated" indicates that this simulation has been done before and that it is not the current simulation (S6); 3. Sentence patterns like "Figure 2/3/4 shows..." indicate the analytic conclusion can be drawn from those prescribed figures (S7, 8, 9, and 16); 4. The simple present tense shows the analysis is given to the current simulation. 5. The passive voice shows the objectiveness of the writers.

continued

S	Original Sentences	Component Analyses
S4–S18	sliding surface (Figs. 4a and 4b, respectively). (S17) The norm of the second sliding surface <u>is driven</u> to zero at $t = t_F^*$ and <u>maintained</u> within a <u>prescribed</u> tolerance for the rest of the flight. (S18) The effect of the rate of convergence <u>can be clearly seen</u> in Fig. 4a, in which the guidance parameter influences the shape of the sliding surface norm.	
S19–S24	(S19) All cases precisely <u>achieved</u> the <u>desired</u> location with a very low terminal landing velocity, <u>but</u> the performances <u>degraded</u> for lower values of Λ. (S20) Indeed, it <u>is observed</u> that, whereas the three components of the terminal velocity are always less than 0.001,5 m/s for $\Lambda = 2, 4$, and 8, the values of the terminal velocity components are close to 0.02 m/s if $\Lambda = 1.5$. (S21) In this specific case, the rate of convergence of the sliding surface is slower and the MSSG algorithm tends to generate, on average, acceleration commands of lower magnitude during the flight. (S22) For a fixed time of flight, the errors tend to be larger toward the end. (S23) Consequently, the MSSG algorithm tries to compensate by increasing the terminal acceleration command. (S24) Despite such guidance command reaction (see Fig. 3), the accuracy <u>is slightly degraded</u>.	1. The first sentence indicates the result of the above calculation and reasoning. 2. Then with a "but," the author changes his tone. He points out a problem: the performances degraded for lower values of Λ. 3. The reason causing the problem is discussed. (S20~S23) 4. 24 shows the result of the first set of the simulation.

Cited from Furfaro et. al., 2013.

Language Focus:

A. Use of mathematical hypothesis.

The authors of scientific papers usually use the expressions like "Assuming (that)..." (S4 and S5), "Once..." (S12), "Suppose..." to indicate mathematical hypothetic conditions.

B. Reference.

Besides using squared brackets to indicate references, as is shown in the previous example, here in this example, we have noticed another way to refer to reference, that is the expressions like "Figure 2/3/4 shows..." to indicate the analytic conclusion can be drawn from those prescribed figures(S7,8,9, and 16).

5.3 Tips on Writing Methods Section

The Methods section is usually the central part of a scientific research paper. It may cover most of the length of the paper. In order to write a good Methods section, we will list some useful tips to help you present your research perfectly before your readers.

1) Be precise

As it is a careful description of the process of your research, to be precise is the uppermost principle. When your method is new to everyone else, it is rather necessary for you to give a detailed picture of your materials, experimental conditions, research procedures or analytic process.

2) Focus on transfer of tenses

When writing this section, the description of the current research is the most important. Meanwhile, you may deal with the conventional scenario of the similar research and the research conducted before yours. Therefore, different tenses are used to distinguish what you did in the current research, what the present situation is and what has been done before.

A. The simple present tense(一般现在时).
- indicates the conventional scenario in experimental research and investigation study;
- indicates the current research in statistical algorithm study.

B. The simple past tense(一般过去时).
- indicates the current research in experimental research and investigation study. It is seldom used in statistical algorithm study.

C. The present perfect tense(现在完成时).
- indicates the research conducted before yours.

3) Pay attention to the use of passive voice

In the previous examples, there are many underlined expressions, as everyone may have noticed. These underlined parts are the passive voice in different tenses and in different grammatical functions. The use of passive voice is critical in writing a scientific research paper, especially in writing the Methods section, because this section is faced with the description of objective matters or things. To use passive voice in different tenses or in past participle forms can justify your objectiveness toward the research.

A) We can obviously notice that in Generalization & Introduction part of this section, passive can frequently be seen in present perfect tense, i.e. "have been+ past participle" form, as is seen in Example 1, which indicates the experiment or investigation conducted before the current one.

B) In the following parts, i.e. Materials, Methods, and Data-analysis parts, passive voice is used frequently in the simple past tense, i.e. "was/were + past participle" form, which gives the detailed description of the materials or the procedures.

C) Passive voice used with *modal verbs* (情态动词), such as "can/may + be + past participle," or "should + be + past participle" are sometimes seen to indicate a kind of possibility or obligation.

D) Also, a past participle is used before or after a noun to be a modifier, or it is used after the main clause to be an adverbial indicating the condition, time, and so forth.

E) Passive voice in infinitive forms can also be seen here and there, indicating a purpose.

5.4 Use Proper Syntactic Templates as Well as Useful Expressions

1) Generalization or introduction

Our methodology involves...

All the samples were prepared/collected/chosen/investigated from/in...
The instruments of this experiment were produced by/purchased from...
The model was performed/conducted/supplied by...

2) Materials

... was positioned near...
... was parallel/adjacent/normal to...
... was covered/surrounded/connected with...
... was embedded in/mounted on top of...

3) Methods or procedures

... was adopted/assembled/calibrated/divided/immersed/maintained...
... was utilized/simulated/substituted/regulated/transferred/stabilized...

4) Data analysis

To achieve/assess/determine/facilitate/guarantee/minimize/prevent...
... with the intention of validating/overcoming/establishing/improving...
As proposed/illustrated/described by...
... is almost/essentially/largely/practically the same with...

5.5 Summary

We discussed the major components of a Methods section and analyzed some examples from different journals, the purpose of which is to enhance the readers' awareness of the four basic components in writing the Methods section. Some tips are given and some syntactic templates and useful expressions are listed to help the readers to write a proper Methods section in their scientific research paper.

Tasks

Task 1
Directions: Read two Methods sections in your research field and identify their basic components.

Task 2
Directions: Choose one of the two Methods sections and translate it into Chinese and then write down the useful expressions including its syntactic templates.

Task 3
Directions: Write one Methods section according to the Instructions of your target journal.

Chapter 6
How to Write Results

What impress readers most in the Results section are the tables and graphs. Someone may argue that since the tables and graphs are sometimes self-explanatory with the help of clear captions or footnotes, there seems no necessity of writing a Results section. Seemingly, such an argument makes sense. However, if you are aware of the sensible reasons proposed by Glasman-Deal (2011: 92), you would not have doubted about the necessity of having a Results section. Her reasons are listed as follows:

➢ Some of your results may be more interesting or significant than others, and it is difficult to communicate this in a table or graph.

➢ It is essential to relate your results to the aim(s) of the research.

➢ In some cases, you may want to offer background information to explain why a particular result occurred, or to compare your results with those of other researchers.

➢ Your results may be problematic; perhaps some experiments were not fully successful and you want to suggest possible reasons for this.

➢ You must communicate your own understanding and interpretation of the results to your readers. Results do not speak for themselves.

The Results section is believed to be the core of a research paper, because it is the *departure point* (出发点) of any conclusions drawn in the following Discussion section.

6.1 Basic Components

Results are sometimes presented separately from the Discussion, sometimes combined in a single Results and Discussion section. For the sake of convenience and clarity, we discuss only the writing of Results here.

Though the Results sections of the published papers may vary from discipline to discipline, from journal to journal, the fundamental function of the Results is still the same, that is, to present data. According to Day and Gastel (2007), the Results section usually consists of two major elements, which are general description of the experiments and presentation of representative data. However, as for how to write the Results section, we believe the 4-component model, put forward by Glasman-Deal (2010:123) and displayed in Table 6.1, can give you a better understanding.

Table 6.1 A 4-component model of Results

1	Revisiting the research aim/existing research Revisiting methodology General overview of results
2	Invitation to view results Specific/key results in detail, with or without explanations Comparisons with results in other researches Comparison/s with model predictions
3	Problems with results
4	Possible implications of results

Very likely, the Results of some published research papers do not include every information element suggested in the model. Some papers may simply go directly to the results at the beginning of the Results without a recall to the research aim or methods, often by inviting readers to look at one of the figures or tables. Some papers make no mention of the problems with results at all. However different they are, to get your manuscript published, you had better familiarize yourself with the components required in your target journal beforehand and follow the requirements in the process of writing.

6.2 Specific Analyses

Many Results sections in their original versions usually have more than a thousand words, which may turn out to be an obstacle for a simple and clear demonstration of its structure. Obviously, a shorter Results section may be easy for the reader to read and for the writer of this chapter to demonstrate the organizational structure of the Results. Therefore, here an abridged Results section is presented for analysis.

Please read the example Results (Table 6.2) very carefully to see whether it includes all the components suggested by the aforementioned 4-component model of Results. To make the task easier for you, the abridged example Results below has been divided into eight parts. What you should do is just to figure out what the writer is writing about in each part.

Table 6.2 The example Results and its component analyses

S	Original Sentences	Component Analyses
S1–S3	(S1) Using our massively parallel magnetic tweezer system, precise, small tensile forces were applied to collagen molecules in the presence of collagenase (see Materials and Methods section). (S2) The experimental series were divided into three categories (see Figure 3): "zero force" (Brownian tether forces ~ 0.06 pN, based on refs 28, 33); "low force" [averaging 3.6±1.1pN (s.d.)]; and "high force" [averaging 9.4±1.3pN (s.d.)]. (S3) The forces were achieved by changing the magnet stack heights to ∞, 2.6 mm or 1.1mm above the surface of the glass.	The writers revisit the details of the Methods.

continued

S	Original Sentences	Component Analyses
S4–S5	(S4) Examination of the energetics of the force application in the range chosen suggests a sharp increase in the stiffness of the collagen monomer from the low to the high force while there is a smaller difference between the zero force and the low force (see Figure 3). (S5) Because of the rapid stiffness increase, we expected to see the major effect of force on the stability of the molecule to occur between the low and high force experiments.	The writers present a general overview of the results.
S6	(S6) The compiled experimental results are shown in Figure 4.	The writers give an invitation to view the results.
S7	(S7) The continuously extracted data from the high and low force runs generally show a classical exponential decay indicating the physics of the enzymatic cleavage process is governed by the law of equally probable events.	The writers report specific results in detail, with explanation.
S8	(S8) This is also consistent with the product formation equation (eq. 25 in ref. 29) and permits direct extraction of the enzymatic cutting rate by fitting the data with e-kt. [29]	The writers compare their results with those in other researches.
S25–S31	(S25) For the 9.4 pN force curve, the slope change occurs at approximately 40s. (S26) Fitting the fast decay regime of the curve gives a kcat of 0.011/s ($r2=1.0$) while the fitting of the slow decay regime gives 0.005/s ($r2=0.984$). (S27) The 3.6 pN force slope change occurs much more quickly (10s into the experiment) and yields kcat values of 0.028/s ($r2=0.99$; fast decay regime) and 0.005/s ($r2=0.998$; slow decay regime). (S28) It is important to note that both the high and low force experiments show similar ejection rates in their slow decay regimes (which we assume comprises ejections of beads with single tethers). (S29) This suggests that the effect of force does not alter the enzyme cutting rate significantly between 3.6 and 9.4 pN for single collagen tethers. (S30) The data obtained from the 0.0 pN experimental series, where the beads were collagen-tethered but unloaded, show a 10-fold increase in the rate of enzymatic digestion (kcat$\approx$0.05/s; $r2=0.97$) relative to the low and high force series (kcat$\approx$0.005/s). (S31) The 0.0 pN enzymatic activity value is comparable to that found when unloaded rhCol1 is degraded in free solution (kcat$\approx$0.06/s).	The writers make comparisons among the three sets of data, describing the relations (i.e. similarity, difference) among them.
S32–S34	(S32) This correlation confirms that our molecular mechanochemical assay accurately captures standard collagen/enzyme kinetics when collagen is tethered to the SPM but unloaded. (S33) These curves are created from the pooled experimental data and fit to one master curve. (S34) This method gives us the most conservative rate of bead ejection.	The writers point out the achievement of the proposed experimental methods.

S	Original Sentences	Component Analyses
S35	(S35) Individual experiment results were used to compare all force states for statistical significance, and it was found that the zero force and control curve were statistically different from each other and every force curve, while the 3.6 and 9.3 pN curves were not statistically different from each other.	The writers make a summary of the key results, echoing the prediction of results.
Cited from Camp, et.al., 2011.		

Language focus:

(1) Location sentences.

The element of **"invitation to view results"** in the Glasman-Deal's 4-component model of Results is to help readers to locate the figure(s) or table(s) where the data can be found. In real practice, such message (e.g. Table 1, Figure 3) is usually stated separately from or combined with data discussion. Furthermore, when combined with data discussion, the location information is either integrated into the sentence or placed in the brackets.

This example Results has altogether four sentences which include the location information as listed below. Let us exam whether or not the location information is combined with discussion on data, and whether the location information is integrated into the sentence or placed in the brackets.

S1 *The experimental series were divided into three categories (see Figure 3): "zero force" (Brownian tether forces ~ 0.06 pN, based on refs 28, 33); "low force" [averaging 3.6±1.1pN (s. d.)]; and "high force" [averaging 9.4±1.3pN (s. d.)].*

S2 *Examination of the energetics of the force application in the range chosen suggests a sharp increase in the stiffness of the collagen monomer from the low to the high force while there is a smaller difference between the zero force and the low force (see Figure 3).*

S3 *The compiled experimental results are shown in Figure 4.*

S4 *In Figure 4, the data from the high load/no enzyme control show that there exists a small percentage of noncollagenase induced ejections and that the population of tethered beads decays at a rate of 0.000,1/s (r2=0.99) over the course of the experiment.*

The analysis results of the location information in the above four sentences are shown in Table 6.3 below.

Table 6.3 The analysis results of location information

Items	Integrated into the sentence	Presented in the brackets
Separately	S3	/
Combined with discussion on data	S4	S1; S2

Interestingly, only one separate location sentence (S3) occurs among the four sentences. In contrast, the rest three sentences (S1, S2 and S4) combine the location information with discussion on data. Moreover, as for the place of the location information in the sentence, S1 and S2 place the location information in the brackets while S4 integrates it into the sentence.

It seems that the authors of the example Results prefer to combine the location information with discussion on data, particularly presenting the location information in the brackets. Probably, such choices are mainly due to personal writing preferences. However, it is not deniable that such choices have their unique advantages. For example, by placing the location information in the brackets, the flow of discussion is smooth, without being interrupted by the insertion of the location information in the sentence. Besides, using this style and the separate location style interchangeably can help to add diversity to the sentence patterns.

(2) Specific/key results in detail.

Reporting quantitative results contains 3 basic elements of information: location of the data, technical description of the data and comment on the data. The following two sentences from the example Results serve as a good example in case.

S6 *The compiled experimental results are shown in Figure 4.*

S7 *The continuously extracted data from the high and low force runs generally show a classical exponential decay indicating the physics of the enzymatic cleavage process is governed by the law of equally probable events.*

It can be seen that S6 indicates the location of the data being discussed whereas the main clause of S7 "The continuously extracted data from the high and low force runs generally show a classical exponential decay" technically describes the data, and the remaining part "indicating the physics of the enzymatic cleavage process is governed by the law of equally probable events" makes a comment on the data.

(3) Problems with results.

No discussion of "Problems with results" can be found in the example Results. Actually, it is also true for many published papers. One reason for its absence is that some authors worry that readers will doubt their results if they honestly mention problems in the results.

However, Glasman-Deal warns that "failing to mention a problem suggests that you aren't sufficiently expert to be aware of it, and this has a negative effect on your professional authority." (2010:121) He argues that it is favorable to admit problems in the results. On one hand, "it shows you to be fully in control of your research and able to evaluate it clearly" (2010:121); on the other hand, "it provides you with an essential element for the Discussion/Conclusion: directions or suggestions for future research." (2010:121)

Though it is acceptable to admit problems in the Results, the Discussion should be worded carefully to "minimize its importance if you can, and suggest possible reasons for the problem/offer a solution" (Glasman-Deal, 2010:121). For this purpose, three techniques can be used, which are

① attributing the limitations to the fact that existing knowledge is unable to resolve the encountered problem,

② deliberately avoiding the word "limitation,"

③ showing that other authors have also experienced similar problems.

Since the example Results makes no mention of its "problems with results," an example from another paper(Ren and Karimi, 2012:281) is presented to illustrate how the troublesome task can be performed tactfully without compromising the reliability of the results. Please find out which of the aforementioned three techniques are used to minimize the seriousness of the problem with results.

E. g. *The accuracy in our experiments is mainly influenced by failure in differentiating two sides of a road by only using stand-alone GPS data*, *so the experimental performance was not as good as those reported by Quddus, et. al. (2007)*, *where differential GPS (DGPS) and DR sensor were used in the experiments. We realize that most mismatched points in our route occur on sidewalks of narrow roads* **due to GPS accuracy limitation.** *Since* **the GPS receiver used in our experiments is a stand-alone GPS with less accuracy than DGPS**, *its accuracy limitation is one of the most important influencing factors results in mismatched points.*

It can be found in the above example that the authors try to moderate the problem of less satisfying accuracy by attributing it to the accuracy limitation of the GPS receiver used in the experiments (i. e. *due to GPS accuracy limitation*). That is, the problem lies with the innate deficiency of the experimental equipment, which is something unavoidable to any other researchers who have used or will use this device.

6.3 Tips on Writing a Good Results

What types of data should be presented in the Results section? How should the data be presented, in the table, graph or text? How should I comment on my tables and figures in the text? Questions like these should be carefully and thoroughly concerned before you start writing the Results section. This is due to the fact that the nature of data presented and the way data are presented in the tables or graphs and interpreted in the text by the author will influence the way readers perceive the results.

1) How to present data in the text

Generally speaking, the presentation of data should be **concise**, **clear** and **logical**.

A. Be concise.

Only representative data should be presented in the Results section. It is not acceptable to simply transfer experimental data and observed facts to the Results part without discrimination. Do not reiterate in words of what is already apparent to the reader from examination of the figures and tables. You just need to point out the key result or trend that the figure or table conveys. Do not repeat word for word the *legend* (图表说明) to your figures and tables within the main text.

B. Be clear.

The language you use to describe your results is as powerful as the tables and graphs themselves. If you do not make explicit comments on your results in the text, there is a risk that the reader may perceive them differently from you. If you lose the chance of guiding the reader to understand your data in a way that you expect, your interpretation of the results in the Discussion section may encounter suspicion or even disagreement from your reader.

C. Be logical.

If the Results section can be written in a few paragraphs, each paragraph should just cover one idea. When substantial information needs to be presented in this part, it is a good choice to split the Results section into several divisions with the help of subheadings. Whichever way you use to present the data, you should remember to "organize them in a sequence that highlights the answers to the aims, *hypotheses*(假设) or questions" (Wallwork, 2011: 233) you put in the Introduction of the paper.

2) When to use a table or figure

The results of many experiments can be presented either as tables or as graphs, but how to present the data should not be a choice randomly or carelessly made. Actually, the choice depends not only on the nature of the data but also on the point you want to deliver to your readers. As Day and Gastel suggest, "your choice might relate to whether you want to impart to readers exact *numerical*(数字的) values or simply a picture of the trend or shape of the data." (2007: 93)

Cargill and O'Connor (2013: 26) come up with a more detailed list of situations (Table 6.4) when either tables or figures are most useful to highlight data.

Table 6.4 The choice between tables or figures for data presentation

Tables	Recording data (raw or processed data) Explaining calculations or showing components of calculated data Showing the actual data values and their precision Allowing multiple comparisons between elements in many directions
Figures	Showing an overall trend Comprehension of the story through "shape" rather than the actual numbers Allowing simple comparisons between only a few elements

3) How to design effective tables and graphs

Since data presentation styles vary with disciplines and journals, it is always the first and the best choice to refer to the *Instruction to Contributors* (or *Instructions for Authors*), or articles in recent issues of your target journals.

4) How to arrange tabular material

Traditionally, tables are made up of both horizontal and vertical lines (Table 6.5). Presently, tables with three horizontal lines (but no vertical lines) are preferable and widely used (Table 6.6).

Table 6.5 Main structure parameters of the vehicle model

Parameters	Units	Values
m	kg	1,000
I_z	$kg \cdot m^2$	600
L	m	2

continued

Parameters	Units	Values
B	m	2
a	m	0.8
b	m	1.2
k_{xf}	N/rad	40,000
k_{xr}	N/rad	40,000
K_{yf}	N/rad	30,000

Table 6.6 Main structure parameters of the vehicle model

Parameters	Units	Values
m	kg	1,000
I_z	kg·m²	600
L	m	2
B	m	2
a	m	0.8
b	m	1.2
k_{xf}	N/rad	40,000
k_{xr}	N/rad	40,000
K_{yf}	N/rad	30,000

Since a table has both left-right and up-down dimensions, the data can be presented either horizontally or vertically. But very often the data are organized vertically so that the like elements read down, not across, as shown in the tables on the previous page.

The vertically-organized format is preferable, "because it allows the reader to grasp the information more easily, and it is more compact and thus less expensive to print." (Day and Gastel, 2007:89)

It can also be seen that the words in the second column are *lined up* (对齐) on the left while the numbers in the third column are lined up on the right (Table 6.6). It should be noted that some journals require the numbers to be lined up on the *decimal point* (小数点).

5) How to prepare figures

Compared with tables, the author has more choices when designing a figure, because several types of figures are available like pie charts and bar charts. However, you should be aware that different types of figures allow you to emphasize different qualities of the data as suggested by Cargill and O'Connor (2013: 27) below.

Pie charts: highlight proportions of a total or whole

Column/bar charts: compare the values of different categories when they are independent of each other

Line charts: display a sequence of variables in time or space or other dependent relationships (e. g. change over time)

Radar charts: present categories which are not directly comparable

In the publishing process, the journal may shrink your figures to fit the journal page or column width so that the words and symbols may become less distinct if their font sizes are not chosen carefully.

If your paper contains two or more graphs that are most meaningful when viewed together, consider placing the graphs above and below each other rather than side by side so as to minimize the impact of reduction.

The importance of the figure legend has not received sufficient attention from authors, especially inexperienced ones. Very likely, a student writer may come up with a figure legend as short and simple as "Comparison results" (Fig. 6.1). Such a figure legend can hardly live up to the requirements to an effective legend, that is, "the reader should not need to read the rest of the text to understand them (figure legends)." (Day and Gastel, 2007: 31)

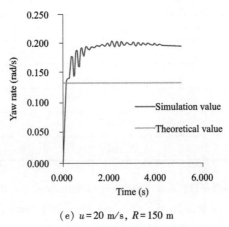

(e) $u = 20$ m/s, $R = 150$ m

Fig. 6.1 Comparison results

6.4 Syntactic Templates in Writing the Results Section

1) Invitation to view results

➢ The compiled experimental results **are reported in Figure 4**.

➢ **Shown in Fig. 2 are** the number of images, in sorted order, collected for each of 9,163 camera configurations.

➢ **Figure 3 illustrates** the findings of the spatial time activity modeling.

2) Specific/key results in detail

➢ For the 9.4 pN force curve, the slope change **occurs** at approximately 40 s.

➢ The sensitivity of SEM **is much higher than** light microscopy.

➢ The control spectra **show slight overlap with** minimally labeled cells but no overlap with maximally labeled cells.

3) Comparison with other results

➢ Distributions **are almost identical** in both cases.

➢ This **is also consistent with** the product formation equation (eq. 25 in ref. 29) and permits direct extraction of the enzymatic cutting rate by fitting the data with e-kt. [29]

➢ This **is in contrast to** our earlier report of an identical ionomer structure with a PF6- counterion.

4) Possible implications of results

➢ This **suggests** that the effect of force does not alter the enzyme cutting rate significantly between 3.6 and 9.4 pN for single collagen tethers.

➢ It **is evident** that the level of labeling varies from cell to cell.

➢ The continuously extracted data from the high and low force runs generally show a classical exponential decay **indicating** the physics of the enzymatic cleavage process is governed by the law of equally probable events.

6.5 Summary

This chapter presents a 4-component model of the Results section established by Glasman-Deal. Though in real practice some published papers do not include every information element as suggested in the model, the presentation of the significant data of the research work is always an indispensable part of this section. Besides, it should be noted that in some journals the Results section is very often combined in a single Results and Discussion section, instead of being presented separately from the Discussion.

Tasks

Task 1

Directions: Read two Results sections in your research field and identify whether they cover all the components as suggested by the 4-component model, and how they are arranged.

Task 2

Directions: Go through as many Results sections in your research field as possible. Identify statements of "problems with results" and figure out which of the three techniques presented in the subsection **"Problems with results"** are used to minimize the seriousness of the problem. If you discover techniques that differ from the three suggested ones, please record your findings for reference when writing your own paper.

Task 3

Directions: Write one Results section according to the Instructions for your target journal.

Chapter 7
How to Write Discussion

In terms of general function, if the Results section is to present the data, the Discussion section is to discuss and reveal any possible relations between the data. In terms of content and organization, the Introduction and the Discussion should function as a pair. Glasman-Deal (2010: 154) claims that "many of elements of the Introduction occur again in the Discussions/Conclusions in (approximately) reverse order."

7.1 Basic Components

The structure of the Discussion section may vary from discipline to discipline, from journal to journal. Just like the 4-component Results model illustrated in the preceding chapter, Glasman-Deal (2010:179-180) again proposed a 4-component model for the writing of Discussion section (Table 7.1).

Table 7.1 A 4-component model of Discussion

1	Revisiting previous sections Summarizing/Revisiting general or key results
2	Mapping (relationship to existing research) (描绘与现存研究的关联性)
3	Achievement/Contribution Refining (提炼) the implications
4	Limitations Current and future work Applications

Brief as this model, some points (e.g. revisiting previous sections, mapping) are barely comprehensible for rookie scientific writers. For this reason, another 6-element model (Cargill and O'Connor, 2013:60) is introduced here to give the inexperienced writer a relatively easy template to follow as well as a clear picture of the organization of the Discussion section. According to Cargill and O'Connor, except for the first and the last elements, the remaining elements are usually repeatedly covered for each group of results that is discussed.

(1) A reference to the **main purpose or hypothesis** of the study, or a **summary of the main**

activity of the study.

(2) A restatement or **review of the most important findings**, generally in order of their significance, including

a. Whether they **support the original hypothesis**, or how they contribute to the main activity of the study, to answering the research questions, or to meeting the research objectives; and

b. Whether they **agree with** the findings of other researchers.

(3) **Explanations for the findings, supported** by references to relevant literature, and/or speculations about the findings, also supported by literature citation.

(4) **Limitations** of the study that restrict the extent to which the findings can be generalized beyond the study conditions.

(5) **Implications** of the study (generalizations from the results; what the results mean in the context of the broader field).

(6) **Recommendations** for **future research** and/or practical **applications.**

Actually, the two above mentioned models are almost identical except for some minor differences in the presentation order of certain information. For example, in the 4-component model, "limitations" of the results are not discussed until the contribution and implications of the study have been presented; in contrast, in the 6-element model, "limitations" of the results are advanced ahead of the discussion of the implications of the study. Such differences in the two models suggest that no single model can serve as a universal model for papers in different research fields. Therefore, if possible, you had better try to work out a model of Discussion based on your extensive reading of the published papers in your target journals, so that your articles can meet the requirements.

7.2 Specific Analyses

Read the example Discussion section below. It is extracted from the same research article as the example of Results section. Such an arrangement is purposely made to provide the readers with a clear and complete picture of the writing of the Results and Discussion sections as a whole, because some research articles have a combined Results and Discussion section.

Please read the example Discussion (Table 7.2) very carefully to see whether it includes all the components suggested by the two aforementioned models. To make the task easier for you, the example Discussion below has been divided into seven parts. What you should do is just to figure out what the writer is writing about in each part.

Table 7.2 Example Discussion and its component analyses

S	Original Sentences	Component Analyses
S1	(S1) The results are consistent with the hypothesis that mechanical strain is a potent regulator of collagen/enzyme interaction.	The writers refer back to the hypothesis and give a general overview of the results. (Element 1)

continued

S	Original Sentences	Component Analyses
S2	(S2) We found that the presence of a small mechanical force (3.6 pN) applied to collagen molecular tethers profoundly enhances their resistance to enzymatic cleavage, but the resistance does not proportionally increase even when the force is raised by more than a factor of 2 to 9.4 pN.	The writers refer to specific results of the present study. (Element 2)
S3–S6	(S3) Though the experiment has been conducted using recombinant collagen in isolation, the results suggest that the strain-stabilizing effect which has been found in both native tissue[18-21] and reconstituted collagen gels[22, 23] can be attributed, at least in part, to factors which occur at the molecular level. (S4) Figure 5 compares the present data to those extracted from the recent investigation of Zareian et al.[21] which examined the effect of tensile mechanical force on the enzymatic degradation rate of native bovine corneal tissue strips. (S5) Both the single molecule and native tissue data follow similar trends whereby increasing the load decreases the rate of enzyme conversion of the collagen. (S6) The data imply that the mechanochemical signature of the collagen/enzyme pair is reflected in the buildup of tissue hierarchy.[35]	The writers compare the findings of the present study with those of other researches, revealing their similarity. (Element 3)
S7	(S7) However, there are substantial differences between native tissue and our single molecule assay which could render the collagenolytic reaction in tissue less sensitive to force than exposed, loaded single collagen monomers.	The writers state the contribution of the present study by highlighting the difference between the present work and another study. (Elements 3 & 5)
S8	(S8) There is too little known about enzyme degradation in the whole native tissue for us to speculate about why there is a difference in the sensitivity to force, so we choose to leave the question open for future investigations to resolve.	The writers mention the limitations of the present study and suggest a direction for future work. (Elements 4 & 6)
S9–S10	(S9) The most striking aspect of the data indicates that tensile mechanical loads cause a rapid switch in the state of the collagen monomer converting it from "enzyme-susceptible" to "enzyme-resistant" at relatively low force values. (S10) In addition, because bacterial collagenase aggressively attacks collagen at multiple sites[36] we conclude that tensile mechanical strain generally enhances collagen stability.	The writers explain key results of the present study. (Element 2)

S	Original Sentences	Component Analyses
S11–S13	(S11) A more general stability enhancement of the triple helix (rather than a specific effect such as changes in enzyme binding site conformation) is consistent with data obtained from investigations of collagen thermal denaturation. (S12) Such studies have shown collagen to be more resistant to thermal denaturation when under tensile strain and when packed into fibrils. [37-39] (S13) The stability enhancement mechanism is not known but has been attributed to decreases in the configurational entropy of the monomers. [40]	The writers identify agreement between the present research and other studies. (Element 3)
Cited from Camp, *et. al.*, 2011.		

From the above table, it can be seen that the example Discussion, particularly its first paragraph expectedly includes every information element as described by the 6-element model. What's more, for each group of results that is discussed, the elements 2 and 3 are usually repeatedly covered just as Cargill and O'Connor claim.

Language focus:

(1) Adjectives in statements of contribution.

When it comes to stressing the contribution of your work, normal flat phrases cannot serve the purpose well. If you really want your contribution to be seen and appreciated, it is time to use adjectives, either qualitative (e.g. remarkable, exciting, invaluable, undeniable) or quantitative (huge, massive) ones. However, frequent use of this kind of language should be avoided because it will eventually compromise the effect.

The following two sentences are statements of contribution from the example Discussion (Table 7.2).

(S7) *However, there are **substantial** differences between native tissue and our single molecule assay which could render the collagenolytic reaction in tissue less sensitive to force than exposed, loaded single collagen monomers.*

(S8) *The most **striking** aspect of the data indicates that tensile mechanical loads cause a rapid switch in the state of the collagen monomer converting it from "enzyme-susceptible" to "enzyme-resistant" at relatively low force values.* (S10) *In addition, because bacterial collagenase aggressively attacks collagen at multiple sites we conclude that tensile mechanical strain generally enhances collagen stability.*

It can be concluded that "**substantial**" in S7 is a quantitative adjective, describing how big the differences are between the native tissue and the single molecule assay, and "**striking**" belongs to the qualitative adjective.

(2) "Limitations" in the Discussion section.

Seeing these words, you may wonder how "limitations" in Discussion section is different from "problems with results" in the Results section. If we have already admitted the problems with our results, is it necessary to repeat it here in the Discussion? Good questioning!

Actually, the statement of "limitations" here in the Discussion section is meant to "point out a direction for future work" (Glasman-Deal, 2010:178). The "limitations" should be ones "which can be addressed in future work, rather than limitations which are inherent to your research field or problems which are unlikely to be solved in the near future." (Glasman-Deal, 2010:178)

It should be noted that when stating the "limitations," we should word them tactfully to lessen their negative impact, just as what we do with "problems with results" in the Results section.

The sentence below is Sentence 8 from the example Discussion (Table 7.2).

There is too little known about enzyme degradation in the whole native tissue for us to speculate about why there is a difference in the sensitivity to force, so we choose to leave the question open for future investigations to resolve.

It manages to reduce the negative impact of the limitations by: ① attributing the limitations to the fact that existing knowledge is unable to resolve the encountered problem "speculate about why there is a difference in the sensitivity to force"; ② deliberately avoiding the word "limitation."

(3) Hedging (运用模糊限制语) and strength of claim (论断的力度).

After presenting experimental data in the Results part, writers are supposed to state their interpretations of the experimental results in the Discussion section. That is to say, they are supposed to make claims about things that they believe have been proved in their experiments.

It has been found that when making claims in the Discussion, native writers of English tend to use a special group of words which lower the certainty of the statement. As in the sentences below, when stating the reason for the extinction of the dinosaur (恐龙), the writer puts the statement as:

E.g. 1: *The extinction of the dinosaur **may** be due to the collision of a small planet onto the earth.*

E.g. 2: *The extinction of the dinosaur is **probably** due to the collision of a small planet onto the earth.*

Can you see the similarity between the two highlighted words, "may" and "probably"? They are similar in meaning, both of them expressing a lower sense of certainty. When we are not quite sure about what we say, we tend to use such kind of words. Therefore, *hedges* (模糊限制语) are expressions of uncertainty, and hedging is the technique of expressing ourselves in a less certain way.

Can you imagine what will happen if we remove "may" and "probably" from the sentences? Will it make a difference? Without "may" and "probably" the sentence becomes:

E.g. 3: *The extinction of the dinosaur **is** due to the collision of a small planet onto the earth.*

With "may" and "probably," the first and the second sentences sound 50%-70% and 80%-90% certain about what it claims, respectively. In contrast, without "may," the third one sounds 100% certain about the reason for the extinction of the dinosaur. However, the third statement is very likely to bring about criticism on the author from native reviewers and other researchers, because in their eyes, the author was too confident about what he/she claimed.

It is risky to be over-confident here, because it is very likely that as science and technology develop, what we believe to be true may be modified or even overturned in the future. This constitutes/leads to the first reason for using hedges in the discussion of experimental data. That is, **"Hedging" would protect the author from any future contrasting findings or conclusions by other authors** (Wallwork, 2011: 137).

If you are still not convinced of the necessity of hedging claims, some research findings on the similarities and differences in the use of hedges between English-native writers and Chinese writers in their English scientific papers will give you further insights into this issue. It has been found that English-native writers tend to use more hedges than Chinese writers. For example, Yu and Qin pointed out that **Native writers use nearly 28% more hedges than Chinese writers in English academic writing.** And this is the second reason for using hedging devices in our claims. Researchers believe that this discrepancy in the use of hedges is rooted in culture. But just as a saying goes, do as the Romans do. When we write in English, and especially, when we want to publish our paper in an international English journal, we'd better follow what the English reviewers expect us to do. However, in real practice, when discussing the data, some Chinese writers appear to be too arrogant and assertive about their conclusions. Though some of them do not mean to do so, they may be wronged by their imprudent/improper use of language.

Now that you know the necessity of using hedges in your discussion of results, some of you may feel worried about how to express your points confidently but not too assertively in your writing. Actually, learning to use hedges in your discussion of results may be less difficult than you imagine. Wallwork (2011: 138) sums up three ways to lower the certainty of the claim "to a degree that most referees would consider to be a more appropriate level of assertiveness, confidence and certainty."

① **Toning down verbs.**

Verbs like is/are, mean, demonstrate, prove and manifest make strong claims of one's findings. If expressions such as **may, would, seem to, appear to, tend to** are added ahead of them, it will reduce the power of them.

② **Toning down adjectives and adverbs.**

Some adjectives and adverbs have very strong tones, for example, innovative, novel, extremely important, very significant, clear(ly), evident(ly), definite(ly), undoubtedly, and undeniably. To sound modest and tentative, expressions like **somewhat, to a certain extent, relatively, likely** and **possibly** should be used.

③ **Toning down the level of probability.**

Modal verbs (e.g. **can, may, might, could**) and some nouns (e.g. **likelihood, probability, possibility**), adjectives (e.g. **possible, probable**) and adverbs (e.g. **possibly, probably, likely, unlikely**) which can express different degrees of probability can also help to reduce the strength of claim.

Besides the above presented three ways of hedging, some verbs expressing a lower degree of certainty are also frequently used in the science writing, for instance, **appear, assume, suggest** and **imply**.

The authors of the example Discussion also use some hedges in their important claims as expected.

E. g. 4: *Though the experiment has been conducted using recombinant collagen in isolation, the results* **suggest** *that the strain-stabilizing effect which has been found in both native tissue and reconstituted collagen gels* **can** *be attributed, at least in part, to factors which occur at the molecular level.* (S3)

It can be seen that two hedges are used in this sentence, one is a verb hedge (**suggest**) in the main clause, and the other is a model verb (**can**) in the subordinate clause. The two hedges make the strength of the statement weaker than if only one hedge is used.

Other sentences containing hedges from the example Discussion are listed below.

E. g. 5: *The data* **imply** *that the mechanochemical signature of the collagen/enzyme pair is reflected in the buildup of tissue hierarchy.* (S6)

E. g. 6: *However, there are substantial differences between native tissue and our single molecule assay which* **could** *render the collagenolytic reaction in tissue less sensitive to force than exposed, loaded single collagen monomers.* (S7)

E. g. 7: *The most striking aspect of the data* **indicates** *that tensile mechanical loads cause a rapid switch in the state of the collagen monomer converting it from "enzyme-susceptible" to "enzymeresistant" at relatively low force values.* (S9)

It can be found that in each of the statements above, only one hedge is used, which are "imply," "could" and "indicate," respectively. Besides, hedges are not used very frequently throughout the example Discussion. It should be noted that proper use of hedges may help you sound modest as well as confident, but overuse of hedges may make your points become "vague and unsure." (Wallwork, 2011:146)

7.3　Tips on Writing a Good Discussion

According to Hu, et. al. (2010:118), many submitted research articles are rejected due to "a faulty discussion" or because "the true meaning of the data in a paper may be completely obscured by the interpretation presented in the discussion." That is to say, even if your results are significant, a poorly written Discussion section will have a destructive impact on the quality of your paper.

How can we achieve a good Discussion? Brief, direct and logical expression of ideas is definitely indispensable. Besides, the following tips may help you to improve your writing of the Discussion section.

1) Interpret your results rather than repeat them

This is determined by the major functions of the Discussion section. As mentioned at the very beginning of this chapter, the Discussion section is to discuss and reveal any possible relations between the data whereas the Results section is to present the data. Besides, repetition of Results is not only meaningless but also boring. Therefore, you should explicitly present how you view your results, so that your readers may arrive at the same conclusions as yours.

2) Show how your study is related to other researches explicitly

Generally speaking, if a research article does not show its relationship with existing research in the Discussion section, it is probably because: ① this article touches upon a research topic that has not been studied by any other researchers, or ② the writer of this article does not make an exhaustive investigation about what has been published on the concerned topic. However, actually in most cases a research is associated with other studies in one way or another. Besides, only when you show how your research is related to other studies can your readers understand the contribution of your paper to the research field.

3) Discuss the theoretical implications and practical applications of your work sensibly

To impress the reviewers or readers, inexperienced writers may tend to overstate the achievements of their research. Their attempt can hardly influence the judgment of the reviewers and experienced researchers. On the contrary, such attempt may lead the reviewers to form a negative opinion of them and their papers. Therefore, being honest and sensible may be the best choice when it comes to stating the contribution of your research.

7.4 Syntactic Templates in Writing the Discussion Section

1) Relationship to existing research

➢ A more general stability enhancement of the triple helix (rather than a specific effect such as changes in enzyme binding site conformation) **is consistent with** data obtained from investigations of collagen thermal denaturation.

➢ The results here also **relate to** the proposed "B-type" blinking process deduced from recent single-particle spectroelectrochemical measurements on CdSe/CdS QDs.

➢ The data presented here **are** thus **broadly consistent with** the interpretation of electrochemically suppressed blinking proposed in ref. 51.

2) Achievement/Contribution

➢ **The most striking aspect of the data indicates** that tensile mechanical loads cause a rapid switch in the state of the collagen monomer converting it from "enzyme-susceptible" to "enzymeresistant" at relatively low force values.

➢ Furthermore, the method **differs from previous approaches** in that the missing values are related to the decision problem rather than to the data distribution.

➢ Nevertheless, the demonstration here of large electrobrightening in ZnSe QDs **reveals a promising and accessible approach to** characterization of redox-active surface traps in QDs, **with important implications for improving** our understanding and control of QD photophysics.

3) Limitations

➢ **There is too little known about** enzyme degradation in the whole native tissue for us to speculate about why there is a difference in the sensitivity to force, so we choose to leave the question open for future investigations to resolve.

➢ **Unfortunately**, because of the film geometry of the DMA experiments imposed by the limited quantity of materials, mechanical properties at higher temperatures **were unattainable**.

➢ At this time, however, these interesting microscopic issues **remain unresolved**.

4) Future work

➢ The power of our forensic analysis lies in the ability to acquire signatures from a wide variety of cameras and cell phones. This poses significant challenges as new cameras and cell phones are constantly released. **We expect to continue** building our database of images and camera information **in order to keep up with these continual changes**.

➢ **Future work may involve** calculating substituted aromatic polarizabilities, and using them in multiparameter equations instead of Mr values, to predict Ebind.

➢ There is too little known about enzyme degradation in the whole native tissue for us to speculate about why there is a difference in the sensitivity to force, so **we choose to leave the question open for future investigations to resolve**.

7.5 Summary

This chapter presents two models of the Discussion section, a 4-component model by Glasman-Deal and a 6-element model by Cargill and O'Connor. Notably, the 6-element model is likely to be more reader-friendly to the inexperienced writer, because it provides a relatively easy writing template for writers to follow. As for the language features, the issue of hedging is particularly expounded to stress the importance of using hedges properly in the claims in the Discussion.

Tasks

Task 1

Directions: Read two Discussion sections in your research field and identify whether they cover all the information elements as suggested by the 6-element model, and how they are arranged.

Task 2

Directions: Go through two Discussion sections in your research field and look for cases of hedging. Count the frequency of each hedge to find out which expressions are more frequently used.

Task 3

Directions: Write one Discussion section according to the Instructions of your target journal.

Chapter 8
How to Write Conclusion

Once you have constructed Discussion, you will close your scientific paper by drafting the Conclusion. Documenting Conclusions needs to read guidelines of your target journals as different scientific journals usually require different types of Conclusions. Some feature a Discussion and a Conclusion, while others combine the two as one. For the sake of consistency, here we will follow the separate mode by focusing on the Conclusion alone. Chapter 8 will thus begin by addressing the functions and components of Conclusion and then exemplify strategies for creating a well-crafted Conclusion.

8.1 Functions and Components of Conclusion

The author substantiates Conclusion with the expected information driven by featured functions. Here the functions will precede the sorted basic components accordingly.

1) Functions of Conclusion

Any scientific paper has to make full and complete sense. Thus the Conclusion is generally where you summarize your well-reasoned arguments and provide a final perspective on your thesis statement. These expectations will be met by the particular functions or purposes of Conclusion.

However, Conclusion does not only restate your paper. Beyond a restatement of your thesis, effective Conclusions often fulfill the other three primary functions.

A) To summarize the research.

B) To elaborate the well-reasoned argument.

C) To emphasize the significance by relating the application or/to future work.

To show how these functions to be met, we will analyze how two well-structured Conclusions unfold the components from the basic analysis to the extended one.

2) Components of Conclusion

Different from the Introduction, the Conclusion moves readers from the world of your paper back to their own world (Rosen, Behrens and Branscomb, 1999). Thus following these intrinsic functions, the Conclusion informatively comprises the restatement of the research findings, elaboration of the thesis statement, and ideas for further research and significance of the research, which roughly match **summary**, **elaboration** and **climax** (Van Rys, Meyer and Sebranek, 2011).

Essential to any Conclusion is the **summary**, which does not merely repeat the thesis but restates the presented argument. Such a **summary** has dual functions. One is to announce to readers subtly that the line of argumentation is finished and the other is to motivate the readers to keep reading the evaluation of the significance.

The expanded summary will then lead to the **elaboration**, which develops what has been left open after the summary. The coming elaboration involves relating fairly back to the context in Introduction, elaborating the paper's significance and/or its implications for future research. Thus your argument will suggest a list of recommendations for future research. In the elaboration, the authors often reveal their particular values or concerns by connecting their papers with the broader concerns of their practical applications or conceptual understanding of some issue. Such concerns will reach the **climax** when the last sentence encapsulates the elaborated ideas.

However, it is not enough to know about the three basic components because an effective Conclusion requires articulating them reasonably. Thus you need to unfold the components in the specific-general pattern by streamlining every sentence into a brief, yet clear Conclusion. Sample 1 will demonstrate how self-conscious writers frame the components compactly.

8.2　Specific Analyses

1) Sample Conclusion 1 (Table 8.1 and Table 8.2)

Table 8.1　Sample Conclusion 1

S	Original Sentences
S1	We **have proposed** a wide-ranging equation of state and a reaction rate law that describes PBX-9502.
S2	The equation of state was *calibrated*(校正,调整) to fit the shock and product Hugoniot curves that fit experimental data.
S3	The reaction law model was calibrated to fit Dn-k and shock initiation data.
S4	Particle velocity wave forms from EM *gauge*(测量,估计) measurements are predicted with excellent agreement.
S5	The forms and procedures are simple enough that they can be modified for other explosives.
S6	Parameters for condensed explosives with more ideal performance such as PBX-9501, PETN, PBXN-9, and *nitromethane*(硝基甲烷) have also been developed, and the results confirm the straightforward application of the constitutive forms and procedures presented here for use with other explosives.
S7	Current work involves **applications** of this model to *detonation*(爆炸/声;爆发) *diffraction*(衍射) and detonation-material interaction problems.
S8	**Further work** including stability analysis, sensitivity of the model to numerical resolution once implemented in simulation codes, and improvements including dead-pressing effects will be investigated.
Cited from Wescott, Stewart, and Davis, 2005.	

Table 8.2 Basic analysis of sample Conclusion 1

S	Description of Discourse Functions	Elements Covered
S1	The writer restates the thesis paper by integrating the most essential information.	Summary
S2–S6	The writer then elaborates on the well-reasoned argument of the research.	Elaboration
S7	The writer justifies the significance of the research by relating it to its applications.	Elaboration
S8	The writer points forward or outward to show further research.	Climax

Extended analyses:

This Conclusion opens by restating the main points compactly. The restatement unfolds itself by: (a) **elaborating** the whole research, and then (b) **connecting** the paper's findings to a larger context **and suggesting** the applications and future work by justifying the significance of the research. Thus such a logically consistent Conclusion will more likely impress the readers with open minds to see their originality.

2) Sample 2 (Table 8.3 and Table 8.4)

Table 8.3 Sample Conclusion 2

S	Original Sentences
S1	We have considered the problem of optimally allocating the leader task between pairs of selfish *UAVs* (无人机 Unmanned Aerial Vehicles) flying in formation.
S2	Formulated as a nonlinear program, our problem poses two distinct challenges: given a fixed number of leader switches, determining the optimal leader allocation and finding the optimal number of leader switches.
S3	We showed that, when switching the lead has no cost, the optimal value can be obtained via a *convex* (凸面的;凸面体) program and designed the cost realization algorithm to determine an optimal cooperation-inducing leader allocation.
S4	In the costly switching case, we restricted the feasible set of allocations to mimic the structure of the solutions provided by this policy.
S5	The resulting restriction has the same optimal value and, for a fixed number of leader switches, is convex.
S6	We also unveiled a *quasiconvexity* (拟凸性)—like property of the optimal value as a function of the number of switches and designed the binary search algorithm to find the optimal number in *logarithmic* (对数的) time.
S7	Future work will include extensions to formations of more than two UAVs, scenarios with obstacle avoidance/no-fly zones, and problems where UAVs bargain over the possibility of joining in formation.
Cited from Richert and Cortés, 2013.	

Table 8.4 Basic analyses of sample Conclusion 2

S	Cohesive Markers	Interplay Between Logic and Structure
S1	We have considered the **problem** of optimally allocating the leader task between pairs of selfish UAVs flying in formation.	The first sentence integrates the research topic into a well-constructed structure. The subject "we" and the active voice show us that the writers are quite confident about their research.
S2	Formulated as a nonlinear program, our **problem** poses two distinct challenges: given a fixed number of leader **switches**, determining the optimal leader allocation and finding the optimal number of leader switches.	S2 starts with the "problem" as the topic, which changes from the New information in S1 into Given information here. Such coreference helps readers build up the logic chain of the ideas. Moreover, S2 also lays out "two challenges" as the new information in the Stress position, signaling to the reader the two challenges will be elaborated in the following sentences.
S3	We showed that, **when switching the lead** has **no cost**, the optimal value can be obtained via a convex program and designed the cost realization algorithm to **determine an optimal cooperation-inducing leader allocation**.	Following the well-reasoned argument as posed in S2, S3 presents the major results by specifying "when switching the lead has no cost," which mirrors back to the key words in S2. Furthermore, the Stress position introduces the "program" and describes the "algorithm to **determine an optimal cooperation-inducing leader allocation**."
S4	In the **costly switching** case, we restricted the feasible set of allocations to mimic the structure of the solutions provided by this policy.	S4 elaborates the other costly switching case in contrast to the no cost case in S3. The writer reports what they did "to mimic the structure of the solutions" in the Stress position.
S5	The resulting restriction has the same optimal value and, for a fixed number of leader switches, is convex.	S5 summarizes the results obtained in the two cases. Although plain, they sound conclusive and solid. Here the resulting restriction in the Topic position mirrors back to the key word in previous sentences.
S6	We **also** unveiled a quasiconvexity-like property of the optimal value as a function of the number of switches and designed the binary search algorithm to find the optimal number in logarithmic time.	The writer reminds the readers that they are moving to a new idea by using "Also" explicitly. S6 further elaborates the key words in S2–S5 by "unveiling a quasiconvexity-like property..." and "designing the binary search algorithm..." The writer keeps justifying the research in a well-reasoned manner.
S7	**Future work** will include extensions to formations of more than two UAVs, scenarios with obstacle avoidance/no-fly zones, and problems where UAVs bargain over the possibility of joining in formation.	"Future work" explicitly moves the reader to the Climax of the Conclusion. The writer outlines the three most important future directions. That is why S7 is both brief and informative, leaving the most important messages to the readers.

Extended analyses:

This is a well-written Conclusion because it is cohesive in the language and coherent in its logic. The cohesiveness and coherence build up implicitly or explicitly through transitions between sentences. Sentences hang well together with the flow of ideas that runs naturally, following the rhetorical principles of Given-New information, Repetition of key words and connectives as highlighted.

Here in this sample, the Given-New coreference of information works well for creating a coherent and unified Conclusion as implicitly as possible. Readers will thus better understand the Conclusion by following the interplay between the sentence structure and its meaning because the authors respect their expectation by refining something familiar and down to something new interchangeably.

Logically, this well-reasoned Conclusion will make even a less alert reader understand the meaning either explicitly stated or just implied because the authors have streamlined the flow of ideas analytically. They unfold their ideas through the summary and further research, thus spelling out what is meant to help readers draw inferences. However, creating such a powerful Conclusion mainly has to depend on the authors' capacity of articulating their ideas convincingly and concisely, on condition that the authors have polished the ideas very reasonably and rationally. Thus the following section will coach you how to articulate a well-crafted Conclusion.

8.3 Strategies for Creating Effective Conclusions

What counts as an effective Conclusion varies from one to another. While no one can offer a single right way immediately ready, here we are to address some guidelines accompanied by sample analyses.

1) Implicitly restate your thesis/position

The worst is to simply copy your thesis and paste it at the end of your paper while the smart is to restate the argument in highly integrated expressions to present your logically-consistent argument. This integrative effect requires refining both the wording and constructing sentences compactly. For example, in Sample 2, "*We have considered the problem of optimally allocating the leader task between pairs of selfish UAVs flying in formation,*" the authors, restating the thesis implicitly, communicate the complex ideas in a style that is readable but still complex.

Moreover, an implicit restatement means to open the sentences with an empty phrase and get to the point directly. When transitioning Discussion to Conclusion, you have to remove such signals as "In conclusion" (Table 8.5) because conclusive language itself already literally sounds like a conclusion, which often anticipates no awkward signal words to overemphasize the conclusiveness. If you are truly endeavoring to provide significance to your paper, you may provide all the proper transitions where they are due.

Table 8.5 Explicit signals for Conclusion

It is important to keep in mind that ...
And, Therefore, Finally, Last but not least...
To summarize, In conclusion, In summary, In closing, In the final analysis...

Omit these phrases. "In conclusion" alike can effectively help lead the audience by showing them the roadmap for an oral presentation, but in writing they are usually perceived redundant or overly mechanical. Thus Conclusion which is supposedly synthetic does not encourage such explicit signals or gap-fillers.

2) Never frame Conclusion with firstly, secondly and finally

Such chronological organization sometimes finds its overuse in Chinese English. It does work well for what happens in a temporal sequence, but it will distract or misdirect readers because Conclusion does not necessarily coincide with the chronological arrangement logically. Conclusion grows with its intrinsic logic flow, which supposedly develops through the corresponding transition means to indicate the argumentation such as opposition, agreement or cause-effect, contrast or comparison. Using these transitions properly will help articulate your ideas faithfully and engage the readers effectively, only when the language form perfectly interplays with its corresponding logic/meaning.

3) Avoid unreadable flow of ideas

Unreadable flow of ideas describes a cluster in which sentences loosely sprawl with the logic implicitly or subtly unreadable. Such low readability prevents the ideas from flowing naturally so that readers will not feel right about the logic and content as you do otherwise. Moreover, as Conclusion is self-contained, it rises around its internal logic flow, which supposedly highly integrates both the logic and its language (as language actually substantiates logic). To demonstrate how sentences in Conclusion better fit together compactly, here we will analyze and revise a sample by considering its language and logic (Table 8.6 to Table 8.8).

Table 8.6 Sample Conclusion 3

S	Original Sentences
S1	In this work, guaranteed closed-loop performance under economic model predictive control (EMPC) over finite-time and infinite-time operation of a nonlinear system was considered.
S2	Owing to the dependence of prediction horizon length on closed-loop economic performance with EMPC, a two-layer Lyapunov-based EMPC was proposed to effectively divide dynamic optimization and feedback control tasks and thus, ease the computational burden of the lower (feedback) layer LEMPC responsible for process control (stability and robustness).
S3	In the proposed two-layer LEMPC structure, performance and terminal constraints are generated by an auxiliary LMPC and then, imposed on the LEMPC optimization problems leading to guaranteed closed-loop economic performance improvement under LEMPC over the auxiliary LMPC.

continued

S	Original Sentences
S4	The two-layer LEMPC structure was applied to a chemical process network to demonstrate the closed-loop stability, performance, robustness, and computational efficiency properties of the proposed two-layer EMPC structure.
* Cited from Ellis and Christofides, 2014.	

Table 8.7 Analyses of sample Conclusion 3 for flow of ideas

S	Analyses
S1	The first sentence is not well-balanced, opening with a long subject and ending by "was considered." This arrangement will leave the readers wondering where the verb is because they usually expect the verb to follow the subject immediately. Moreover, the main idea should be in the "given" position. Shift "given" information closer to the beginnings of your sentences when you can, so that the topic is clear (Gopen and Swan, 1990). In addition, beginning with "In this work," unnecessary and less informative, feels redundant and less straightforward.
S2	In the second sentence, "divide" and "ease" imply causation, which is hidden in "and" relationship. Such loose organization of ideas shows the implicit logic.
S3	The readers have difficulty immediately understanding "... optimization problems leading to guaranteed closed-loop economic performance improvement under LEMPC over the auxiliary LMPC" because the complex wording has buried the most intended message. It will make sense more immediately when we streamline every phrase to highlight the subordination structure.
S4	In "The two-layer LEMPC structure was applied to a chemical process network to demonstrate the closed-loop stability, performance, robustness, and computational efficiency properties of the proposed two-layer EMPC structure," "Applied" and "demonstrate" can be more straightforward if replaced by more accurate, concrete and important verbs. The sentence will make better sense with more compelling and clearer logic.

Table 8.8 Revised sample Conclusion 3

S	Revisions
S1	This paper has examined the guaranteed closed-loop performance under the economic model predictive control (EMPC) over finite-time and infinite-time operation of a nonlinear system.
S2	Owing to the dependence of prediction horizon length on the closed-loop economic performance with EMPC, a two-layer Lyapunov-based EMPC was proposed to effectively divide dynamic optimization and feedback control task, thus easing the computational burden of the lower (feedback) layer LEMPC responsible for process control (stability and robustness).
S3	In the LEMPC structure, performance and terminal constraints are generated by an auxiliary LMPC and then, imposed on the optimized LEMPC that can improve the closed-loop economic performance under LEMPC over the auxiliary LMPC.

	continued
S	**Revisions**
S4	The two-layer LEMPC structure was tested in a chemical process network and the results showed its closed-loop stability, performance, robustness, and computational efficiency.

In addition to the strategies and avoidances covered above, Conclusion also follows a bunch of other explicit or implicit rules for academic writing. These rules will be fully considered while the following sample Conclusion is revised.

8.4 A Sample Conclusion for Revision (Table 8.9 and Table 8.10)

This section will introduce a 3-step process (Bizup and Williams, 2014) for revising the Conclusion in terms of language and readability.

(1) Diagnose: Identify what makes you feel unclear, indirect or unreadable. They are the candidates for revision.

(2) Analyze: Analyze why they do not feel right as meant otherwise.

(3) Revise: Polish them as they feel clear and concise as you would like to read.

Table 8.9 Sample Conclusion 4

S	Original Sentences
S1	The theory of lossless convexification has been generalized to optimal control problems with mixed non-convex and convex control and state constraints.
S2	This was done by introducing the strongly controllable subspaces and an appropriate maximum principle.
S3	As a consequence, a larger class of non-convex problems can be solved as a convex problem.
S4	The work is significant because this class of problems includes several important applications, and the work has been a foundation for successful flight tests of planetary landing trajectory optimization.

Table 8.10 Suggested revisions of sample Conclusion 4

S	Revisions	Analyses
S1	The theory of lossless convexi-fication has been generalized to optimal control problems with mi-xed non-convex and convex control **constraints,** and state constraints.	"With mixed non-convex and convex control and state constraints" in the original S1 causes ambiguity because several stacking modifiers have swallowed the head, "constraints." The revised version is much clearer by adding "constraints" and a comma.
S2	**This application** was realized by introducing the strongly con-trollable subspaces and an appro-priate maximum principle.	"Done" is vague and informal so we replace it by "realize," which is more specific, formal and descriptive. "This" is so ambiguous that readers will have no clue without referring back to S1. That is why "this" is substantiated by "this application."

continued

S	Revisions	Analyses
S3	**Consequently, this integration** solved a large class of non-convex problems as a convex problem.	"As a consequence" is redundant and can be removed or replaced by "consequently" because the sense of causation is already implied literally. "This integration" refers back to the previous sentence.
S4	**The work has brought this class of problems** important applications, and **laid a foundation** for successful flight tests of planetary landing trajectory optimization.	The significance will sound more compelling if better reasoned and evidenced implicitly because Conclusion helps evaluate its significance, especially those readers who have waded through the discussion but are still asking "So what." Moreover, "been" is rephrased by a more descriptive and important verb, "laid."

The analyses and suggestions for revision have covered rules mentioned in previous chapters, as summarized in the final section.

8.5 Summary

This chapter has addressed the Conclusion's functions substantiated by the basic components, and shared strategies by exemplifying analyses of 4 samples. Here the wrap-up will cover the language, the organization and the logic.

Apart from being accurate, clear and concise, language in Conclusion might highly summarize the well-reasoned argument so that we need to use more abstract words and construct the sentences compactly with complexity. In particular, among the basic components, restating the thesis statement requires more than credible concrete evidence, but a highly-integrated, critical, creative and constructive synopsis. This encompassing expectation demands well-refined wording and complex and well-constructed sentences. When we frustrate readers' expectations, we make them work unnecessarily harder.

Furthermore, the language in Conclusion must be well-argued enough to sound convincing in tone and logic. This compelling effect requires considering the logic flow of your ideas. To articulate your points reasonably and rationally, polish your wording by using proper qualifiers, and expressions that make better sense.

Always keep readers in mind and make sure that they can anticipate your use of language and logic. Once you have gained the interest of your readers, you will manipulate their understanding towards your final word.

Tasks

Task 1

Directions: Go through a Conclusion of an SCI paper in your own field. Underline the basic components of the Conclusion.

Task 2

Directions: Now read it again and think about how sentences hang together. Circle the transitions between sentences and underline their Given/New information, references, repetition of key words and connectives.

Task 3

Directions: Finally, reflect on what counts as an effective Conclusion in terms of language, organization and content.

Chapter 9
How to Revise a Scientific Manuscript

Before you submit your writing to a key journal in your research field, doing dozens of revisions cannot be emphasized more. Although revision could be painstaking and sometimes frustrating (when getting rejected), you have to learn that many good writers survived by rewriting over and over. For example, Rowling, the author of *Harry Potter* series, was rejected by 12 different publishers until she got attention from the Bloomsbury Publishing Company. Hemingway continued rewriting the ending to *A Farewell to Arms* thirty-nine times in order to "*get the words right*" (写得准确到位) (Wyrick, 2008: 91). There is no *rule of thumb* (经验规则) for the times of revision needed before publishing, but usually dedicated writers polish their manuscripts from 10 times to 150 times.

9.1　The Definition of Revision

Revising is reading and rewriting one's own work, much in the same way as *straightening up house* (收拾房间): to clean up unnecessary items and put things in right order. However, it is more than mechanical work, it also involves psychological process, a thinking process which enables you to review your draft as a critical reader. It is a daunting task to become one's own reader, because you have to *tailor your draft* (调整初稿) with a principle of effectiveness, question the significance of what you have done, and occasionally get over the reluctance of deleting what you treasured "beautifully written" passages. Undoubtedly big decisions have to be made when it comes to adding an entirely new section, deleting irrelevant statements, or reorganizing your arguments to develop and emphasize your ideas, but it is worthwhile doing so. Any novice writers tend to assume "once a sentence is written, it is *gold or carved in diamonds* (像嵌在珠宝里的金饰永久不变)" (a quote by Hauber, cited from Powell, 2010), yet this is not often true as major or minor revisions are quite common in peer-reviewed journals.

Revising is also rethinking, a process frequently occurring before drafting, during drafting, between versions of drafts, and at the end of several drafts. During this process, the writer is not only required to aim at the overall structure, the style and clarity, but also the smallest elements, such as a misspelled word, or an ambiguous phrase.

9.2　The Peer Review and Revision

The competitive publishing circle requires high-quality research papers that catch readers' attention immediately; therefore the editor-in-chief or editorial staff are the first *batch of gatekeepers* (一批守门员), who *brush it off* (断然拒绝) when the title or abstract does not fit squarely in their scope of interest. If the manuscript is on the *cutting edge* (前沿领域) or is of sufficient quality to be sent for peer review, chances are the manuscript will be examined by two (or more) anonymous peer reviewers. They, in a few weeks or months, recommend or reject your manuscript. Quite often a rejection is frustrating and some experts suggest no more revisions are necessary if the manuscript was rejected three times. But exceptions do exist, especially when the anonymous reviewers are unfamiliar with your research field. On other occasions, it is *laudable* (值得称颂的) that they do not reject your manuscript directly, which indicates a proper revision will make it publishable. Keeping in mind the flow of peer review (Figure 9.1) may be of some help.

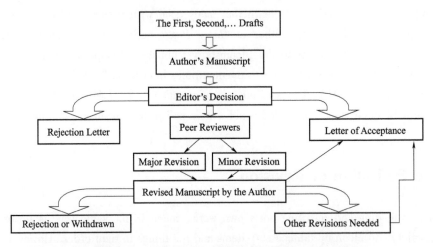

Figure 9.1 The peer review and revision process

Table 9.1 shows a sample of Rejection Letter issued by an editor, who states the reason of rejection and kindly encourages the author to have another attempt.

Table 9.1 A sample of rejection letter

A Sample of Rejection Letter	Analyses
Your manuscript has been reviewed by <u>three experts</u> in the field whose comments can be found at the end of this message or attached as a PDF (for Reviewer 2). The reviewers offer mixed recommendations <u>ranging from "Minor Revisions" to "Not appropriate for *Langmuir*(期刊名称)"</u>. Your recent J. Phys. Chem. B publication reduces the novelty of this manuscript to some degree, although the crystal structure is a new feature. Nevertheless, <u>I do not think this is sufficient to warrant publication in Langmuir,</u> which appeals to a broad audience. Therefore, I am following Reviewer 2's recommendation that the manuscript be declined and <u>instead be submitted to a more specialized journal</u>. I am sorry to disappoint you, especially since your Soft Matter manuscript was also declined.	*The number of referees* *Contradictory comments* *Decline* *Suggestion of another attempt*

Chapter 9　How to Revise a Scientific Manuscript

continued

A Sample of Rejection Letter	Analyses
In considering this manuscript as well as the Soft Matter manuscript and your published work in this area, it might be beneficial to reconsider how you decide to put results together for manuscripts. You are clearly doing a lot of work with these cyclodextrin-threaded polymers, which I personally find to be quite interesting. However, you might have better outcomes with reviewers if you considered combining analyses of different cyclodextrins within the same manuscript. This might allow you to draw broader conclusions from your experimental results that would be more satisfying and convincing to reviewers.	*Suggestion of major revisions* *Encouragement* *Future possibilities*

Two more samples are given below in Table 9.2 and Table 9.3, one being Major Revisions and the other Minor Revisions.

Table 9.2　A sample of reviewers' comments on major revisions

A Sample of Reviewers' Comments on Major Revisions	Analyses
Thank you for submitting your manuscript for publication in *The Journal of Physical Chemistry*. It has been examined by expert reviewers who have concluded that the work is of potential interest to the readership of *The Journal of Physical Chemistry*; however, it appears that a major revision, possibly followed by further reviewer evaluation, will be needed prior to its further consideration for publication. Please see the enclosed reviewers' reports for details regarding the requested changes and/or additions.	*Value of the study* *Requirement for Major Revision*
Reviewer 1	
Recommendation: This paper is publishable subject to minor revisions noted. Further review is not needed.	*Requirement for Minor Revision*
Comments: In the present work, authors have prepared novel polyrotaxane-based triblock copolymers. Different techniques have been used to characterize their structures. Movements of the b-CD wheels along the central axle have been observed due to changes in the temperature and the solvent.	*Creativity of the study*
Reviewer 2	
Recommendation: In my opinion, this paper can be published after minor revisions.	
Comments: 1. Authors use a lot of abbreviations in the manuscript. In order to clarify the comprehension to readers, they might write a list of abbreviations with their meanings.	*Clarification of terms needed*
2. Page 25: Authors must explain, or clarify, how they obtain the number of ×× molecules trapped into a polyrotaxane.	*Clarification of methodology needed*
3. English must be revised.	*Language use*
...	

continued

A Sample of Reviewers' Comments on Major Revisions	Analyses
Reviewer 3 Recommendation: This paper is <u>not recommended</u> because it does not provide new physical insights. Comments: This is a confusing paper. <u>The rationale for constructs design is not explained</u> and the novelty or interest of the results are <u>not very convincing</u>, compared to the previous work published in the literature (work of Ooya and co-workers, references 30-32, 38).	*A negative comment* *Justification needed* *Little contribution*

Table 9.3 A sample of reviewers' comments on minor revisions

A Sample of Reviewers' Comments on Minor Revisions	Analyses

Manuscript ID PY-ART-11-2010-000360 entitled "Dual Thermo-responsive Polyrotaxane-based Triblock Copolymers Synthesized via ATRP of N-isopropylacrylamide Initiated with Self-assemblies of Br End-capped Pluronic F127 with β-Cyclodextrins" which you submitted to *Polymer Chemistry*, has been reviewed. The comments of the reviewer (s) are included at the bottom of this letter.
 	
  <u>The reviewer(s) have recommended publication</u>, but also suggest some minor revisions to your manuscript. Therefore, I invite you to respond to the reviewer(s)' comments and <u>revise your manuscript</u>.
 	
  Referee 1
 	
  Comments to the Author: This paper is <u>a further contribution</u> in the area of main chain polyrotaxanes and cloud point measurements. Please avoid "LCST" and include "cloud point" instead. Finally, the work of Ritter *et. al.* <u>has been fully overseen</u>.
 	
  Referee 2
 	
  Comments to the Author: This paper shows the dual-thermoresponsive behavior of polyrotaxane in aqueous solution. It comes from the thermoresponsive properties of NIPAAm as a capping chain, and of polyrotaxane with the interaction between copolymer and cyclodextrins. Although it can be expected well from the molecular structure, <u>new interesting phenomena are presented with enough experimental data and discussions</u>. Particularly, a change of aggregation form from wormlike to vesicle seems to attract much interest. <u>I think that it should be published as it is</u>.
	Acceptance
	
	Sometimes a time-scale will be given, such as one to three months.
	
	Significance
	
	Updating Literature
	
	Helpful Evaluation
	
	Hurrah!

9.3 Three *Myths* (误区) about Revision

Whether you take initiative to revise your draft or you are suggested to revise by referees, the following misconceptions are commonly held and would better be addressed properly. (Wyrick, 2008)

1) I can ignore some contradictory comments given by the referees

As referees have invested time to make your paper better, you are not supposed to ignore any comment they provided. When the comments given by them are contradictory, for example, one reviewer said it's too long and the other said it's too short, you have to clarify any ambiguity and state your rationale of keeping the length as it is. Early researchers are likely to be deterred from critical comments and do not have the guts to negotiate a feasible scheme.

2) Revision is limited to editing and proofreading

Editing refers to corrections of "surface errors," such as spelling mistakes, ungrammaticality, ambiguity and word diction. *Proofreading* (校正) means examining *typos* (打字错误) *and gaffes* (小错) that distort meaning or distract potential readers. Without question, both editing and proofreading are crucial to the accuracy, emphasis and clarity of an essay, but they are only *mole hills in mountains* (山里的鼹鼠丘，比喻小困难). A dedicated revision includes them but not limited to them. When you are focusing on *small fry* (小问题), typing mistakes for example, you would better keep bigger fish in mind, that is, the global changes you have to make. Global changes are your research purpose, focus, organization, development and significance. Highlight your major argument and argue for it.

3) Revision is punishment or busywork

Some early researchers regard comments as a denial to their long-time hardworking or finding quarrel in a straw by experts. Being asked to pay extra revision work is disappointing especially when graduation is approaching and publications are required to get a diploma. Nevertheless, you have to calm down and think positively about what shall be done first and foremost. More often than not, you can discuss helpful criticisms with your supervisors or colleagues, who may share the same experience, on the one hand, and take the opportunity to help you reshape or refine ideas, on the other. The revision process is, in this sense, a reevaluation and reassessment of your work. Did you get your ideas across effectively? Have you *interwoven* (交织、加入) experimental data with your interpretation into an *integral part* (整体)? Did you leave out some counter-arguments?

9.4 Tips to Revise the Draft/Research Paper

Once you have decided to revise your paper, do not put your hands on it right after finishing the draft. Experienced writers will put the draft aside for a day or two, and then read the draft and comments in an objective and systematic way. By being objective, writers do not take any comments personally or painstakingly argue for their views; instead, they converse with sources. By being systematic, writers do not revise an entire draft in one *lump* (一大块), trying to examine all the parts of the draft, from ideas to organization to mechanics, at the same time. Instead, they normally break revision process into six smaller, manageable steps (Table 9.4). The 1-3 steps emphasize macro-structure revision, revision above discourse level, while the 4-6 steps focus on micro-structure revision, such as the revision of particular words and phrases.

Table 9.4 Six steps taken in revision

1	Reconsider the purpose, contributions and the audience	Macro-structure
2	Rethink about the ideas and evidence	
3	Reconsider the organization of the paper	
4	Rethink about the clarity and style	Micro-structure
5	Edit the grammar, punctuation and spelling	
6	Proofread the entire essay	

1) Reconsider the purpose, contributions and the audience

First of all, your paper has to discuss an issue of current concern in the field, which is stated both in the abstract and the introduction. Your abstract should inform the audience what the paper is about and what the major contributions are. It is highly recommended that a succinct description of previous research should be given to frame your argument. (Cited from The Emerald Group Publishing Website) In the following example, the author gives too much irrelevant information in an abstract:

E. g. 1: ... *The purpose of this paper is to describe the experience of a team of academics in the Department of French, School of Modern Languages within the Faculty of Arts, Humanities and Social Studies at the University of ×× in the development of a computer-assisted learning software program.* ...

Comments: Far too wordy, *mock-posh*(假装正式) and giving irrelevant information. Don't use "academics" unless to distinguish from administrators.

Here is an excellent abstract example demonstrating the background, the purpose and the implication of a study.

E. g. 2: *Fishes swim by flapping their tail and other **fins**(鳍). Other sea creatures, such as squid and **salps**(樽海鞘), **eject**(喷射) fluid intermittently as a jet. We discuss the fluid mechanics behind these propulsion mechanisms and show that these animals produce optimal vortex rings, which give the maximum thrust for a given energy input.* ... *An important implication of this paper is that the repetition of **vortex**(旋涡) production is not necessary for an individual vortex to have the optimal characteristics.*

Secondly, the purpose and contributions of your study should be reemphasized in the final paragraph of the introduction section. The following example fails to achieve this due to a lack of novelty and specific contribution.

E. g. 3: *The present paper details the experience of one team of academics in the French Department within the School of Modern Languages which will hopefully have broadly based implications not only for other disciplines but also within other universities as a whole throughout the UK.*

Comments: The paper is of no interest whatsoever to most readers, if any at all. The purpose is badly expressed, along with many inconsistencies. The paper does not achieve the aim.

To achieve maximum readability, writers have to specify their goals and their novelty in the research field. For example, the following excerpt demonstrates this point by using phrases "in order to examine" and "are broadly supportive of."

E. g. 4: *In addition to the theory, we present results from numerical simulations. These are done in order to examine whether the predicted stable flows can arise naturally as a result of the time-dependent evolution. ... The simulations are broadly supportive of the theoretical predictions, although time-dependence can produce exotic and interesting final states.*

Finally, the target audience will be reviewers, colleagues, experts and scientific students both in the research field and in particular interdisciplinary areas. Therefore you have to explain where your study comes from and clarify terms if necessary.

2) Rethink about the ideas and evidence

Once you settle down the purpose of your writing, you have to offer the readers step by step guidance to accept your argument. That is, your methodology and data analysis have to link coherently to your main idea; all the facts and evidence have to argue either for or against your ideas. Here are a few tips that you can bear in mind:

- If you alter the main idea in your revision, think about whether the evidence is still relevant and can further the readers' understanding.
- If you stick to the original idea, make sure that you have a clear relationship between the thesis and each of the major points presented in different sections, such as the discussion.
- If you have included irrelevant or unnecessary evidence, you would better remove them in order not to *derail*(使……走向歧路) your readers' understanding.
- If you provide data and results, consider whether the experimental data are capable of supporting the conclusions drawn.
- If your main ideas are not located as specified in scientific writing instructions (placing research questions at the end of the methodology section, for instance), think about whether readers can easily identify what position you are taking in each part of your discussion. If not, you have to consider how to frame your argument in an understandable way.

Most importantly, do not present data or results as they are, leaving valuable discussion unattended.

E. g. 5: *Disappointingly, eight of the students for one reason or another were obliged to withdraw from the trial. However, the team believes that the results show, in a broad sense, the potential of CALL within Higher Education. The results speak for themselves.*

Comments: Oh, come on! How can "the results speak for themselves"? Aren't you responsible for inducing and interpreting?

3) Reconsider the organization of the paper

Although scientific writing does not encourage creative writing as in literary styles, good writings are more understandable and enjoyable to read. The entire essay takes the IMRaD format, and each paragraph gives a well-organized flow. For early researchers, the "hamburger structure" paragraph will secure the flow of arguments. In each "hamburger structure," a key sentence, supporting sentences and a concluding sentence represent the upper bun, the beef and vegetable, and the lower bun respectively. The point presented in the key sentence has to be explicitly stated, otherwise your ideas might be overlooked or somehow misunderstood. Additionally, all the major ideas have to flow smoothly between and within paragraphs. One of the effective measures is to use sufficient transitional devices to guide the reader along.

In order to structure the entire essay, creating an outline with the key sentences in each paragraph is efficient, as the organization of key sentences will ensure a convincing, clear and well-organized plan is in place. While you are revising, do not be afraid to restructure your drafts. Though restructuring might be time-consuming, it definitely will increase the incidence of getting your work published.

4) Rethink about the clarity and style

As you have revised for the macro-structure, the purpose, ideas and organization, you have to polish sentences and word choice to ensure your readers' complete understanding of your ideas. Effective sentences and word choice help you develop a clear, concise, lively, or emphatic style. Ask yourself the following questions during this process:

- Are there any sentences that run in circles, leading nowhere? How can I make each sentence contain information pertinent to my point? Can I combine it with the sentence that follows or rewrite it with vivid, specific details? By using active verbs rather than passive ones? **For example**, instead of describing your roommate as "a horrible person," you can portray him as "a person who leaves dirty dishes under the bed, sweaty clothes in the closet, and toenail clippings in the sink."

- Are there any sentence fragments, misplaced words, or convoluted phrases that might cause confusion?

- Are there any sentences that are wordy, irrelevant or redundant? Will they cloud the reader's mind to follow my ideas?

- Are there any repetitive or extra-long sentences to weaken your emphatic style? Do offer a variety of sentence lengths and patterns in your prose.

- Are there gaps or omissions in the coverage, data, logic and presentation?

When reading through the manuscript, reviewers are likely to get annoyed by vague or fuzzy language. For example:

- ... it is <u>the belief of the authors</u> of this paper that much C. A. L. L. software has been designed more with presentation than with learning effectiveness in mind. With TAVAC, <u>screen design</u> was a major factor and consequent upon some staff interviews with the Department of ..., it was decided that the simplest screen was one of the principle aims and objectives, even if it meant a certain <u>loss of flexibility</u> in the end user environment.

- **Comments**: The author's belief would be fair enough if it were backed by referenced evidence. The author does not provide evidence of any special screen design in the screenshot. In addition, the author does not state how flexibility was lost.

5) Edit the grammar, punctuation and spelling

Correcting errors in grammar, spelling, diction and punctuation is like waxing and polishing a car to make it shine. (Wyrick, 2008) Some hints might be helpful:

(1) Read aloud.

Reading silently helps you detect incoherent links, but reading aloud allows your ears to hear ungrammatical "chunks" or unintended gaps in sense or sound you may otherwise miss.

(2) Read backwards.

Try to start reading your draft at the end of your essay and work towards the beginning. You do not have to read each sentence word-for-word backwards; just read the draft one sentence at a time. Usually writers engage in editing their drafts from the beginning of their essays, and they tend to begin thinking about the ideas they try to convey rather than concentrating on the task of editing for errors. Reading backwards sentence by sentence makes it less likely to wander away from the editing job at hand.

(3) Use dictionaries.

Apart from learners' or advanced dictionaries, a thesaurus and a collocation dictionary are highly recommended. A thesaurus can be useful if you want to increase the variety of words or avoid monotonous tone. A collocation dictionary can clarify word usage, such as whether it is authentic or foreign use, whether the word contains positive or negative connotation. Without questions, each word has its meaning from the company it goes.

(4) Ask for others' help.

6) Proofread the entire essay

Read your draft for several times and pay attention to the following three things:

(1) Typing errors.

Even the simplest transposing of letters can change the meaning of an entire thought and occasionally leave negative impression in the reviewer's mind.

Imagine the surprise of restaurant owners whose new release instructed them to "Please sing the terms of the agreement." ("sign" is the correct form.)

(2) Submission format.

You have to follow strictly the scientific writing instructions and the format required by a particular journal. Most reviewers appreciate beautifully placed tables and figures, pages that are numbered and ordered clearly, references that are traceable with either MLA style or APA style.

(3) Clean and professional outlook.

If you want the reviewers to take your manuscript seriously, you have to present your work in a scientific and professional manner. Submitting a paper with different word fonts, some informal jokes, or a coffee stain on it has about the same effect as a *blob* of spinach in your teeth — it distracts the audience from hearing what you have to say.

9.5　Resubmission after Revision

When you have finished all the revision work on details in macro-structure and micro-structure, it is time for you to resubmit your manuscript to an academic journal or the journal that sent you suggestions given by anonymous reviewers. Any suggestions, positive or negative, should be considered and replied honestly and politely. The importance of writing a letter in response to suggestions is similar to a thank-you note to those who have offered you help, and you have to let them know how things *pan out*(顺利发展) and *close the loop*(有始有终).

Table 9.5 shows a revision letter sample to express gratitude and acknowledgments.

Table 9.5　A revision letter sample

A Sample of Revision Letter	Analyses
Dear Leanne, <u>It is with excitement</u> that I resubmit to you a revised version of manuscript JSPR-11-308, Dyadic Perceptions of Goals, Conflict Strategies, and Perceived Resolvability in Serial Arguments for the *Journal of Social and Personal Relationships*. <u>Thank you for giving me the opportunity</u> to revise and resubmit this manuscript. In keeping with my last communication with you, I am resubmitting this revision before the agreed upon deadline, March 16, 2012. <u>I appreciate the time and details provided by each reviewer and by you</u> and have incorporated the suggested changes into the manuscript to the best of my ability. The manuscript has certainly <u>benefited from these insightful revision suggestions</u>. I look forward to working with you and the reviewers to move this manuscript closer to publication in the *Journal of Social and Personal Relationships*. <u>I have responded specifically to each suggestion below</u>, beginning with your own. To make the changes easier to identify where necessary, I have numbered them. ...	*Addressing term* *Gratitude* *Gratitude* *Acknowledgements* *Acknowledgements* *Sincerity*

Showing your courtesy and sincerity is one of the two parts in a revision letter, the other being a list of the reviewers' comments and your revisions. In the list you have to quote their comments, respond to the comments, and state what action has been taken. For example:

- *Practical applications — are the implications for practitioners clearly drawn out?*
  Reviewer's comment: Somewhat weak... The implications are primarily for research purposes.
  Response: Looking back at the paper more objectively, we have to admit you are right. We rewrote implications section by adding new implications for managers and elaborating on

already presented implications.
- *Research applications* — *does the article suggest areas for further research?*

  Reviewer's comment: Further research needs to be stated in terms more relevant to that stream.

  Response: This section has been re-written in Discussions and Conclusions.

- *Analytical rigor* — *does the article demonstrate soundness in the way in which it has been argued?*

  Reviewer's comment: Researchers did not establish validity and reliability of their measures. A qualitative analysis of the data is necessary to demonstrate its validity and reliability.

  Response: Fair point. We have attempted to use well-known and tested measures of main variables to ensure reliability and validity of variables in the research. We indicated source of variables and discussed how other studies used them effectively.

Above all, revision work goes hand in hand with the follow-up letter writing; they will reflect what you have contributed to the knowledge base and what a robust study you have done. Revisions are not *on-and-off business* (时断时续的事情); you have to focus on your manuscript for months until every tiny mistake you spotted is wiped off. Remember, systematic revisions will be paid off!

Chapter 10
How to Submit a Manuscript

A manuscript is the work that an author submits to publishers and editors for publication. Publishing scientific articles is a necessary part of success as scientists. Their success is greatly measured by not only the quality and quantity of research output, but also the impact of that research on other research or practice. But each journal has its own *Instructions for Authors*. This chapter focuses on the submission of manuscripts to optimize the acceptance of an article.

10.1 How to Prepare a Manuscript

Basically, good manuscripts are able to communicate a clear and useful message to the readers of a particular journal. In the preparation of a manuscript for submission, the materials are organized into the appropriate scientific format. It is essential that authors prepare the manuscripts exactly according to the journal's *Instructions for Authors*. Authors can optimize their publication success by understanding and meeting the selection criteria of the journal. Other points to consider include where and how to submit the manuscript.

10.2 Where to Submit

1) Decide on the target journal

When the manuscript is completed, the authors need to decide on where to submit the manuscript. The ideal result is that the manuscript can be published in the best possible journal in the relevant research field in the shortest possible time. Keep in mind a hierarchy of journals according to their reputation. Experienced authors generally know what will or will not be accepted in a particular journal. For less experienced authors, they can ask a *mentor* (导师) for help. Good mentors provide advice to less experienced authors as to where and how to submit the manuscript. Such information as the acceptance rates, and the manuscript processing time, and the journal's impact factor may exert influence on your decision of your target journal. Most commonly, the authors' institutions have certain criteria to fulfill a list of recognized journals. Make sure that the journals are in this recognized category. Thus a careful selection of the target journals is the first step of publication.

2) Familiarize with the target journal

After the journal is identified, the authors should familiarize themselves with the target journal.

Read some recent issues of the journal, either online articles or in printed forms, for the purpose of obtaining a general idea of the style of the journal. Focus on the *mission statement* (宗旨) of the journal.

Tips for authors:
- Search on line or ask experienced mentors for suggestion to get a general idea of all the best possible journals relevant to your research.
- Make sure the target journal is in the recognized category of your working institution.
- Read the recent issues of the target journals.
- Pay attention to the style, mission, *impact factor* (影响因子), etc. of the journal.

10.3 What to Submit

1) **Formatting the manuscript**

(1) **Pay attention to publishers' guidelines.**

It is highly recommended that authors follow the target journal's instructions with regard to the formatting of reference citations. Journals vary in their styles, which calls for authors' close attention. Publishers' *guidelines* (指南) for formatting are the most critical resource for authors.

(2) **Ask a native language user to proofread the article.**

For non-native English authors, it is useful to have a native English speaker edit or proofread their manuscripts if they submit to English journals. Authors that are not proficient in the English language should get professional help.

(3) **Self-edit and revise the article repeatedly.**

Apart from help with language from professionals, self-editing, revising, and repeated revising of the manuscript are of great importance. There are two ways to achieve this. The authors can distance themselves from the recently-completed manuscript before reviewing it again. They can also ask an experienced colleague to critique the manuscript. They can present their proposal to a group of faculty members before writing and present the paper to them before submitting the manuscript.

Reading a hard-copy version of a paper helps to pick up presentational shortcomings that might otherwise be missed. Reading a paper through carefully helps to pick up inappropriate structure and section headings, as well as poor development in argumentation, etc.

(4) **Organize the article according to the particular categories of writing.**

Authors should be aware that each type of paper has its own specific features. Different writings serve different purposes and are judged by different criteria. The types of writings include *original article* (研究论文), *review paper* (综述论文), *case report* (个案报告), *technical note* (技术说明), pictorial essay, *commentary* (评论), *editorial* (社论), letter to the editor, other categories, and non-scientific material (Peh and Ng, 2009: 759). Therefore articles should be organized according to the particular categories of writings.

(5) **Check *Instructions for Authors* for the structure requirements.**

The basic structure of an original scientific paper includes introduction, method, result and discussion, which can be summarized by IMRaD. Detailed instructions on how to organize the material into a formal structure may vary from journal to journal and differ among specific paper types, and can usually be found in the journal's *Instructions for Authors*.

(6) **Follow the house style of a particular journal.**

The style of an article can be divided into *house style* (印刷风格) and individual writing style. House style identifies each particular journal. Authors should follow the house style of the journal and pay attention to such elements as spelling, capitalization, units and references as well. Follow journal style guidelines with respect to the inclusion of an abstract, arrangement of *section headings* (段落标题), *referring style* (引用格式), and where tables and figures should be inserted. Remember to include page numbers.

There are some general principles to keep a good individual writing style. The principles include keeping the manuscript short; writing short sentences; using short instead of long words; avoiding *redundancy*(避免冗余); avoiding *figures of speech*(修辞手法), *idiom or jargon* (习语和术语); trying to avoid passive sentences; taking care with *abbreviations*(缩写词), and being careful with *modifying words and phrases* (修饰词和短语) (Peh and Ng, 2009: 760).

2) **Cover letter** (附信, 投稿信)

(1) **What is a cover letter.**

A cover letter is the one you write to the editorial board of the journal when you submit your manuscript. After all the revisions have been approved by each author, it is time to submit the manuscript for review by the target journal. In addition to the manuscript itself, the submitter must be prepared to send the editors an abstract, a list of keywords (if applicable), author information, *disclosure statements* (公开声明), and a cover letter.

(2) **Contents of a cover letter.**

The cover letter includes statements addressing several items. It may disclose the *conflicts of interest*(利益冲突), which include any information about the authors' professional or financial *affiliations*(附属机构) that could be perceived to influence the content of the manuscript. It describes any previous presentation of the material. It states the responsibility of each author for the completion of the paper. By the time you send the article, the editor should know the basic information about the authors and about the article. It provides such details as names, addresses and phone numbers of sources. Some journals may require authors to provide some reviewers for the manuscript.

(3) **Example of a cover letter.**

The following is an example of a cover letter. The author provides information about how the authors can be reached for possible questions. The cover letter is also a good place to provide contact names and numbers, which are often required by editors.

      Zhang San, Ph. D.
      Environment and Architecture School
      Sichuan University

Mozhiqiao, Ring One Road, Chengdu, Sichuan
People's Republic of China
Zip code: 610088
E-mail: abc@yahoo.com.cn

January 5, 2001
Elizabeth Brewster 5
Food Processing 333 Michigan Ave.
Chicago, IL 60612

Dear Dr. Richard,

I enclose a manuscript entitled ..., which I submit for possible publication in the Journal of... Technology. All of the authors agree to the submission of this paper.

The text includes 20 pages, 3 tables, and 2 figures prepared using Microsoft Word Processing 2003 according to the journal's *Instructions for Authors*. We have provided all required supporting documentation.

We respectfully submit that the following individuals would be suitable peer reviewers based on their expertise in the field:

Li Yu, Microbiological Research Institute, Chinese Academy of Agricultural Sciences. E-mail: li_yu@yahoo.cn.

Zhang Jun, Resources and Environmental Protection School, Sichuan Agricultural University. E-mail: zhangjun@yahoo.cn.

Chen Wei, School of Horticulture, China Agricultural University. E-mail: chenwei@yahoo.cn.

In recent years, our research group has had some conflicts with Li Haha and Wang Chuachua in Sichuan Institute of Microbiology in project application. We request that these individuals not be consulted.

We thank you for considering this work and look forward to your response. Please direct all correspondence about this manuscript to me.

Sincerely
Zhang San
E-mail: abc@yahoo.com.cn

The following are all the authors in this paper and their signatures.

Zhang San	×××
Li Si	×××
Wang Wu	×××
Yang Yu	×××

(*Adapted from http://emuch.net/html/201101/2787763.html*)

10.4　How to Submit

1) Use online submission system

Nowadays, most journals adopt an online submission process. With the use of online manuscript submission system, it is much more convenient and speedy to submit and process a manuscript, and to track its progress.

2) Follow the guidelines to upload a manuscript

Perspective authors and reviewers will need to register their affiliations and other details before being guided through the stages in the process of uploading a submission. Typically, authors will receive an email acknowledgement that the article has been successfully submitted.

3) Wait for a response e-mail from editors for acknowledging the receipt of the manuscript

You will receive an e-mail response acknowledging receipt of your manuscript. The response includes a manuscript number which is useful for all further correspondence. If an acknowledgement is not received within one month of submission, the author should contact the journal office.

4) Deal with hard copy submissions with great caution

The author should keep a complete copy of the submitted manuscript. All data and records should be carefully kept as it may be necessary to re-analyze the original data for manuscript revision. Some journals still require manuscripts of hard copy version. For hard copy submissions, the manuscripts and prints should be packaged to minimize any possible damage.

10.5　Understanding Review Process

1) Review process

When you submit an article to a journal, you need to be patient. Firstly, it is the initial response-time for the editor to read the paper in the context of a flow of other manuscripts that need to be dealt with, identify potential reviewers and get their agreement to write a review. Secondly, the editors receive the feedback from reviewers, read the reviews, reread the paper and write a reasoned decision letter. Thirdly, the author takes time to rework the paper and resubmit. Then the editors process the revised paper followed by the second or third (and subsequent) round (s) of edition. The following is a figure concerning the process from paper submission to publication(Figure 10.1).

[*Adapted from Fayolle and Wright (eds.)*, 2014: 28]

2) Peer-review (同行评议)

(1) What is peer review.

Each journal has its own individual process with regard to how peer-reviews are conducted. Some use a *blinded review* (盲审), which means the reviewers' identity is not revealed to the authors. Others use masking to withhold the authors' identity (name, institution) from the

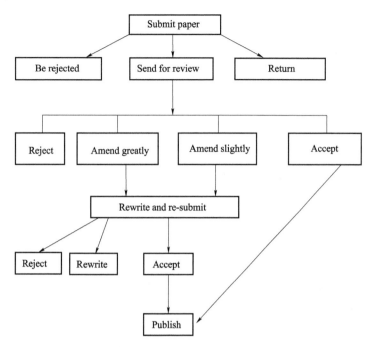

Figure 10.1　Review process

reviewers, which is known as *double-blinding* (双盲) when used in combination with blinding.

B. The function of peer review.

Peer review helps journals to decide whether the focus, novelty, and importance of the research can reach the standard of the journal. It checks that the presentation and style of the content are in accordance with the accepted conventions for production and reader convenience. Peer review advises the authors and the journal editors about the possible ways to improve the manuscript.

3) Review report.

The journal may have asked you to suggest potential reviewers, or the editors may have chosen them from a database of professional networks. Authors will not know who the reviewers are.

The reviewers will be asked to perform the following duties (Margaret and O'Connor, 2009: 77):

- Familiarize with the journal's instructions for reviewers, which are sometimes available on the journal's website.
- Read the manuscript and write a report about the quality of the manuscript and note any problems.
- Recommend any changes that would improve the manuscript.
- Complete an evaluation form about the quality of the manuscript.
- Recommend whether the manuscript should be accepted by the journal or accepted after revisions.
- Return the written report and the evaluation form to the editor.

For example, the *Nature* journal requires that an ideal review should answer the following questions:

- Who will be interested in reading the paper, and why?
- What are the main claims of the paper and how significant are they?
- Is the paper likely to be one of the five most significant papers published in the discipline this year?
- How does the paper stand out from others in its field?
- Are the claims novel? If not, which published papers compromise *novelty*(新颖)?
- Are the claims convincing? If not, what further evidence is needed?
- Are there other experiments or work that would strengthen the paper further?
- How much would further work improve it, and how difficult would this be? Would it take a long time?
- Are the claims appropriately discussed in the context of previous literature?
- If the manuscript is unacceptable, is the study sufficiently promising to encourage the authors to resubmit?
- If the manuscript is unacceptable but promising, what specific work is needed to make it acceptable?

A review report could begin with a short summary of the key findings and value of the manuscript. A good summary benefits the reviewer, editor, and authors. Comments on the manuscript can be arranged under major and minor points. Major comments are generally about the validity of the study, whether study samples were chosen appropriately and whether proper statistical tests were conducted accurately. Minor comments may include clarification of specific sentences, requests for additional data, or grammatical errors.

When the review report is received, the authors can use journal review reports to improve the manuscript and demonstrate to the journal editors how improvements have been made and resubmit the manuscript. It is advisable to address the points in major comments and minor comments one by one.

10.6 Dealing with Rejection

When first submitting his manuscript to the target journal, the author should consider the "fallback" plan in case that the initial submission may not be accepted. Try to revise and/or resubmit as quickly as possible. The author should decide whether the revised manuscript should be submitted to a new journal or to resubmit to the same journal.

10.7 Summary

To conclude, authors should have a clear picture of all the best possible journals before choosing the target journal. Following the target journal's instructions for authors is the most important stepping-stone for the acceptance of the manuscript. Meanwhile, a good manuscript should be constructed in the format that best presents the authors' research results and is in accordance with the house style of the specific journal. If possible, submitting the manuscript via online system is a more efficient and speedy way. A good understanding of the review process is of great help in the process of preparing and submitting a manuscript.

Tasks

Task 1
Directions: Make a list of some highly reputated journals in your research field. Familiarize yourself with the *Instructions for Authors*.

Task 2
Directions: Choose two target journals and compare their similarities and differences in the format requirements. Write or revise your paper according to the format of one of the journals.

Task 3
Directions: Suppose you have finished a scientific paper and expect to get it published in your target journal. Write a cover letter to the editor of the journal.

Chapter 11
International Conference

International conference is one of the most important means of academic communication and top conferences play more positive roles in the latest research than top journals. More and more students of different disciplines are participating in international conference where English is the medium of communication. Adequate knowledge about international conference is the key to a successful communication in the conference. This chapter aims to provide guidelines to those who are planning to participate in an international conference.

11.1 Kinds of Meetings

Thousands of academic meetings are held worldwide every year. However, there are different kinds of meetings with different names such as seminar, colloquium, symposium, conference, forum, and convention, in which you are required to obtain sufficient knowledge so as to choose proper kinds of meetings to participate in.

A meeting is a general term for any kinds of gathering of people for a particular purpose. A *seminar*(研讨会) is a class-like meeting in which one chief speaker gives a lecture on a specific topic or problem while other members have a discussion about the topic. A *colloquium*(高端研讨会) is a large academic seminar like *panel discussion*(小组讨论) attended by certain invited experts. A *conference*(国际会议) is a formal, professional or academic meeting lasting for a few days at the international level. A *symposium*(专题会议) is smaller than a conference but has more specific topics than a conference. There may be several symposiums in a conference. A *forum*(论坛) is a public meeting with a specific theme on public issues on which participants share visions. For example, the theme of 2015 Boao Forum For Asia (BFA) is Asia's New Future: Towards a Community of Common Destiny. A *convention*(年会,例会) is regularly organized by a learned society, a professional association, an academic institution or a non-governmental organization.

11.2 Conference Documents

Conference documents are the most important source of conference information. Main conference documents include Call for Papers, Letter of Invitation and Conference Program.

1) Call for papers

Call for papers (presentations) is the announcement of a conference. It includes information about the conference name, background, objectives, theme, topics and subtopics. It also includes conference dates, *venue* (会议地点), and brief introduction of *keynote speakers* (主旨发言人), as well as the organizing institution, sponsors, abstract (full paper) submission information, presentation format, accommodation and registration, local transportation and contact information.

Sample Call for Papers

Call for Papers
The European Conference on Sustainability,
Energy & the Environment 2015
ECSEE2015
The Thistle Brighton, Brighton, United Kingdom
Thursday, July 9, Sunday, July 12, 2015

Conference Theme: "Power & Sustainability"

Sustainability remains as a powerful ideal that acts as a driving force in relation to the pressing concerns of energy and the environment. Its proponents often protest their lack of power to effect change because of the control systems already in place in the form of tax-evading multinational corporations, banks that support them, and governments that turn a blind eye to malpractice. Is the claim to lack of power as serious as is alleged, or are there other approaches that have not been tried? Advocates of sustainable growth surely have the responsibility to search for alternative avenues and corridors of power. The abolition of slavery in the British Empire, for example, was the work of one man who resisted and confounded the argument that no economy could survive without it. This conference will take the concepts and lenses of power and sustainability as we seek multidisciplinary solutions and work together towards the creation of a sustainable world. We look forward to seeing you (again) in Brighton in 2015.

Submissions are organized into the following thematic streams:
- Environmental Sustainability & Human Consumption: Food and Water, Hunger and Thirst
- Environmental Sustainability & Human Consumption: Waste
- Environmental Sustainability & Environmental Management: Freshwater, Oceans and Seas
- Environmental Sustainability & Environmental Management: Land Use & Misuse
- Environmental Sustainability & Environmental Management: Atmosphere and Air
- Cultural Sustainability: Protecting, Preserving and Conserving
- Economic Sustainability: Environmental Challenges and Economic Growth
- Economic Sustainability: Sustainable Businesses and CSR
- Social Sustainability and Social Justice
- Social Sustainability, War and Peace
- Social Sustainability and Sustainable Living
- Energy: Environmental Degradation

- Energy: Renewable Energy and Environmental Solutions
- Energy: Energy Economics and Ecological Economics

Environmental Sustainability & Human Consumption: Human and Life Sciences

Deadlines

- Abstracts submission: May 15, 2015
- Results of abstract reviews returned to authors: Usually within two weeks of submission
- Full conference registration payment for all presenters: June 15, 2015
- Full paper submission: August 12, 2015

How to Submit

- Register with our online submission system.
- Create your account. Your email address will be used as your username and you will be asked to submit a password.
- Submit your abstract of no more than 250 words, choosing from the presentation formats listed below (Individual, Poster or Virtual).
- Submit well before the submission deadline in order to benefit from Early Bird rates.
- Your proposal will normally be reviewed within two to three weeks after undergoing a double blind peer review. Those who submit near the May 15, 2015 deadline will usually receive results by May 29, 2015.
- If your proposal is accepted you will be invited to register for the conference. Upon payment of the registration fee, you will be sent a confirmation e-mail receipt. (Cited from http://conferencealerts.com)

2) Letter of invitation

Letter of invitation is sent to the participants a few months before the conference. At the beginning of the letter, the addressor first expresses the pleasure to invite the addressee. Then he offers brief information about the conference, such as the full name of the conference, date, venue, website and the purpose of invitation. At the end of the letter, he often expresses the wish for the acceptance of the invitation to the conference.

Sample invitation letter:

iice 2015		iicll 2015
The IAFOR International Conference on Education	iafor	The IAFOR International Conference on Language Learning

International, Intercultural, Interdisciplinary
Crowne Plaza Dubai Festival City, Dubai, United Arab Emirates
Sunday, March 8, 2015-Tuesday, March 10, 2015

Dear Prof....

On behalf of the IAFOR local organizing committee and the IICLL 2015 conference chair,

Professor Sue Jackson, I am pleased to inform you that your proposal, "Project-Based Learning in English for International Conference," has met the accepted international academic standard of *blind peer review* (盲审), and has been accepted for Oral Presentation at IICLL 2015 and the opportunity for your full paper to be published in the official conference proceedings.

The conference will be held in Dubai, UAE, at the InterContinental Festival City Event Center from the evening of Sunday, March 8 through Tuesday, March 10, 2015. The keynote and plenary session will be on Sunday morning and parallel panel sessions begin Sunday afternoon and run for the duration of the conference. Most panels run for 90 minutes, with three presenters per panel, so each presenter has 30 minutes total for presentation and Q&A. For more detailed information about the conference and accommodation, please visit the conference website.

If you can not present for any reason, please notify the conference administration team at <u>iicll@iafor.org</u>. If there is a day that you are unable to present, please contact the administration team at the time of registration. Not everyone can be accommodated with preferential dates and time, so please limit any request of this nature to unavoidable situations.

A PDF of the full Conference Program will be uploaded on the IICLL 2015 website by 2/22/2015. Please check the program at that time to make sure all information pertaining to you is included and correct.

Thank you for participating in The IAFOR International Conference on Language Learning—Dubai 2015. All of us affiliated with the organization aim to make this conference a success.

Your Sincerely,
Signature

Joseph Haldane, Ph. D. (London), F. R. A. S.
Executive Director, IAFOR
iafor
IAFOR, Sakae 1-16-26-201, Naka Ward, Nagoya, Aichi, Japan
460-0008 www.iafor.org

3) Conference program

Conference program is a guide as well as a plan for the participants listing all conference activities at a definite time and place. It usually includes the following information: a letter of welcome written by the conference chairman, activities, date, time, place and people involved. Some conference programs also include the introduction of keynote speakers, all the participants' abstract and the map of conference rooms. Every participant could get a program when registering.

Sample conference program:

The IAFOR International Conference on Language
Learning 2015   Dubai, UAE
Sunday Session II 15:15-16:45

Sunday Session II 15:15-16:45   Room AL Amwaj I
IICLL- Higher Education: Sciences and Life Sciences

Session Chair: Shu-Mey Yu

11887 15:15-15:45
Modeling Heart Cardiac System Using Differential Equation and Matlab
Seifedine Kadry, American University of the Middle East, Kuwait
Aisha Bouhamad, American University of the Middle East, Kuwait
Yousel Al-Hajri American University of the Middle East, Kuwait
7862 15:45-16:15
An Evaluation of Science Lectures' Testing Skills in Tertiary Institutions in Nigeria: A Case Study of Kogi State University, Anyigba
Hassana Oseiwu Ali, Kogi State University, Nigeria
11194 16:15-16:45
University Students' Life Science Critique-Oriented Argumentation Learning Progression
Shu-Mey Yu,"National Taichung University of Education," Taiwan of China

Sunday Session II 15:15-16:45   Room Al Khayma
IICLL-Technology Enhanced and Distance Learning
Session Chair: Anna Toom

6790   15:15-15:45
Tools for Realizing Online Teaching in Higher Education
Cynthia Northington Purdie, William Paterson University, USA
7199 15:45-16:15
Students' Experience with Blended Learning Using a Flipped Classroom Approach
Oogarah-Pratap Brinda, Mauritius Institute of Education, Mauritius
Gungadeen Anuradha Mauritius Institute of Education, Mauritius
6441 16:15-16:45
Students-Enthusiasts in a Virtual Classroom: Their Contribution to the Education Process
Anna Toom Touro College, USA

16:45-17:00
Coffee Break
17:00-18:00
IICLL 2015 Featured Workshop Presentation
Room: AL Baraha 3
Healthy Children Healthy Minds: Helping Children Succeed Now for a Brighter Future
Marcel Lebrun, Plymouth State University, USA

18:30-20:30
Conference Reception

Join fellow delegates for light snacks and a drink at the conference reception. The reception will be held on the Crown Plaza Leisure Deck. This is a great chance to network and meet fellow delegates, which all registered presenters and audience are welcome to attend. Admission is included as part of your registration fee. Please note that alcoholic beverages will be served at this event.

(Cited from http://iafor.org/conference)

11.3 Conference Activities

Activities in a conference involve two kinds of meetings: formal meetings and informal meetings. Formal meetings include Opening Ceremony, Closing Ceremony, *Plenary Session*(全体会议), Parallel Session and Poster Session.

All the participants, government officials, reporters are required to participate in the Opening Ceremony and Closing Ceremony. In a Plenary Session, well-known keynote speakers present their papers and papers about some research subjects of universal significance and general interests are also presented. In Parallel Session, several smaller-scale meetings are held simultaneously. A special topic is arranged for a particular parallel session and participants of a particular research group present their papers. In a Poster Session, a summary of the research paper is posted on a board.

Informal meetings are meetings for free communication like free information exchange and free paper presentation. Visits to famous places, such as research institutes, universities, museums, or historical spots are also informal meetings. Other informal meetings are various social events, such as banquets, parties, excursions, and games.

In addition to formal meetings and informal meetings, tutorial (teaching and consultation), virtual presentations (online presentations), exhibitions, business talks, scientific expedition may be found in some conferences.

11.4 Conference Stage

In the evening of the registration day, a formal reception dinner is held. The next morning, officials, keynote speakers and all the participants present at the opening ceremony. Then the keynote speakers present their papers. Between the opening ceremony and keynote presentations is a coffee break, when conference photos are taken and delicious cakes and nice tea/coffee/fruits are offered. From the afternoon of the opening ceremony day to the day of the closing ceremony, keynote presentations and parallel session presentations take place at the same time. Participants have a variety of presentations to choose to listen to at this stage. The closing ceremony is held on the last day of the presentations. A farewell banquet is served typically on the last night of the conference and all the participants leave the next day.

11.5 Ways to Present Papers

There are six ways to choose to present your paper at a conference: Oral Presentation, Poster Presentation, Virtual Presentation, Workshop Presentation, Panel Presentation and Author Non-Presenter. Oral Presentation is the standard format for presentation and is usually thirty minutes in length including asking and answering questions after the presentation. In a Poster Session papers are posted on boards or walls in a specially separated area and the participants read and discuss with the author about the research in the spare time at the conference. Poster presentations give participants an opportunity to talk about their research with those who are interested in their research.

Virtual Presentation is offered to some presenters who could not go to the conference to present their papers, mainly due to financial and/or political restrictions on travel. Virtual presentations allow authors the same publication opportunities as regular presenters. A Workshop Presentation is a special kind of course led by an experienced practitioner lasting 60 to 90 minutes. It emphasizes group interaction and is featured by problem solving, skills training, or the demonstration of new disciplinary approaches. Many conference workshops become conferences or symposiums after several successes. Panel Presentation is a group presentation lasting for about 90 minutes. A Panel **must** have at least four participants with the chair leader as the primary author, and the other presenters as co-authors. An Author Non-Presenter is an author of a paper who wishes to attend the conference but does not want to present.

11.6 Conference Abstract

Conference abstract is a standalone, self-contained description/representation of the research work and is essentially the "selling point" for your paper helping the conference organizer decide whether they should invite you to the conference or not. A conference abstract is a summary-informative description of your research paper including the following information about your research:

- the research background, purpose and problem statement/hypothesis
- thesis statement
- research methods (procedure/approach/material, data analysis methods/tool)
- results(findings) and discussions
- conclusions(major findings, contributions, implications, applications)

The number of words of most conference abstracts is limited to 250-350 words and no more than 3 paragraphs. To make sure that the reviewer can immediately see the relevance, importance and scientific merit of the abstract, you should make certain English is perfect, avoid more words than required causing readers to skim, and avoid leaving out the summary results or conclusions causing readers to lose interest.

Sample conference abstract:
Annual Bio-medical Research Conference for Minority Students
Nov. 7-10, 2012 SAN Jose, California

This abstract was submitted and accepted to present at the 2010 ABRCMS. It contains all of the components necessary for consideration of acceptance into the conference. Permission to reproduce the submission was obtained from all authors and the mentor.

Understanding Self-efficacy and Well-being in Patients with Schizophrenia

(1) Quality of life in patients with schizophrenia can be adversely affected by factors such as impaired cognitive functioning and other symptoms. However, positive intrapersonal characteristics may offset these factors and improve their well-being. Therefore, identifying positive psychological resource factors is crucial, particularly those that may improve quality of life. (2) This study had two specific aims: 1) to examine the relationship between self-efficacy and well-being, and 2) examine psychosocial factors that are associated with increased self-efficacy. (3) Participants were 62 middle-aged or older participants (Mean age = 50.4, SD = 6.2), with a DSM-IV chart diagnosis of schizophrenia or schizoaffective disorder. Self-efficacy was measured using the Revised Self-efficacy Scale (RSES). Participants' perceived well-being was measured using the Recovery Assessment Scale (RAS). Factors that we anticipated would be associated with self-efficacy were: a) Behavioral Activation, measured using the Behavioral Activation for Depression Scale (BADS), this scale assesses participants' engagement in structured activities; b) depression, measured using the Calgary Depression Scale (CDS); c) social contact, measured using the Lehman Quality of Life Index (QOLI). This scale assesses the frequency with which participants did things with friends, such as attended events outside the home or talked on the phone, etc. We first examined correlations between scores on the self-efficacy scale and those measuring well-being. (4) Significant correlations were found between social self-efficacy and total RAS scores, $r(60) = 0.63$, $p < 0.001$. For the RAS sub-scales, we found significant correlations with personal confidence and hope, $r(60) = 0.62$, $p < 0.001$, willingness to ask for help, $r(60) = 0.37$, $p = 0.004$, goal and success orientation, $r(60) = 0.54$, $p < 0.001$, reliance on others, $r(60) = 0.53$, $p < 0.001$, and not feeling dominated by symptoms, $r(60) = 0.48$, $p < 0.001$. A simultaneous multiple linear regression was then performed. For this analysis, social self-efficacy was the criterion variable, and BA, social contact, and depression were the predictor variables. The model including all variables accounted for 32.3% of the variance in social self-efficacy, $R2 = 0.323$, $F(3,54) = 10.06$, $p < 0.001$. Significant predictor variables included BA ($\beta = 0.277$, $p = 0.019$), social contact ($\beta = 0.282$, $p = 0.016$), and depression ($\beta = -0.341$, $p = 0.003$). Participants' self-efficacy is associated with greater well-being. Also, greater behavioral activation, greater social contact and less depression significantly predict high levels of social self-efficacy. (5) Our data are correlational; therefore, caution should be used when interpreting these effects. However, increasing behavioral activation and social contact via psychosocial interventions may help to increase social self-efficacy and improve quality of life in patients with psychosis.

(Cited from http://abrcms.org)

Chapter 12
How to Deliver Presentations

A presentation is an event at which you describe or explain a new product or idea (Longman Dictionary of Contemporary English, 2004). It can be a talk, speech or monologue. As for academic communication, presentations take place at meetings, conferences, workshops, forums, seminars, symposia and so on. A presentation includes two parts: the delivery of the paper and the question-and-answer session (often shortened as Q & A session). Successful presentations can enhance the understanding of the main ideas of your paper among other professionals and improve your status within the academic circle (Cong and Li, 2007: 109). However, a presenter must have a clear understanding of essential presentation skills and helpful techniques for delivering an effective presentation.

12.1 Features of an Effective Presentation

There has never been a unanimous agreement among scholars about the answers to "What are the features of an effective presentation?" In this chapter, we just choose the following three features to discuss in detail.
1) good preparation
2) clear organization
3) efficient expression

12.2 Tips on an Effective Presentation

1) Good preparation
To make the presentation well-prepared, firstly you should ask yourself the following three questions:
(1) Do the contents match the needs of your audience?
(2) Do you know the materials thoroughly?
(3) Do you use pictures, tables or charts to prompt yourself for certain information?

An important but always overlooked activity in the preparation is to know your audiences' needs since they can be more diverse than the readership of an academic journal. Before starting to write the manuscript, you should have a clear idea about your audience, such as their knowledge level, assumptions and research interests, all of which can help you overcome the stage fright as well as overexcitement in the delivery. Another benefit of knowing your audiences is that you can make some changes in both written and oral expressions to fit their "styles" and to make the delivery smooth.

E. g. 1: *Suppose that you are giving a presentation of SARS'S symptoms, causes and prevention measures to four groups of people: foreign tourists, elementary students, average city residents and medical personnel. Will you use the same written and spoken languages for the same purpose? Absolutely not. Instead you have to investigate their needs, knowledge and interests in the subject, then decide how to arrange your materials. Furthermore, how can you generate the interest among them?*

Think of these questions before you get down to preparing how to present your information so that you are not talking down to, or over the head of your audiences.

Secondly, study the materials thoroughly. Check and double check the accuracy of the information in your manuscript. Spend some time in library, search information on line or simply discuss the contents with fellow scholars to equip yourself well with details and in depth. Otherwise, you will put yourself in a troublesome situation in the question-and-answer session.

E. g. 2: *The following two sentences are from a student's presentation for the thesis defense, which one is better and why?*

a. The Grimm's Fairy Tales was first published in the 19th century.

b. The Grimm's Fairy Tales was first published in 1812.

Both of them are understandable while the first sentence contains the general information "the 19th century" and "1812" in the second sentence is more specific and accurate. Therefore, the second one is preferred.

Thirdly, prepare visuals like tables and figures in your PPT slides for the data presented in graphics are simple and universal. (Day and Gastel, 2007:170) They are interesting, impressive and easy to be compared with. However, you had better write and explain clearly what are the purposes for using the visuals before you go to them and try not to bombard the audience with overloaded tables and charts because too much is as bad as too little.

Now please take a look at E. g. 3, a well-constructed table (Table 12.1) and E. g. 4, an impressive figure (Figure 12.1), both of which are chosen from a student's homework.

Table 12.1 Comparison of HP60Q turbocharger bearing-rotor critical speed

Calculation and Experimental Results	The First Critical Speed (r/min)	The Second Critical Speed (r/min)
D (The calculations without the seal structure)	66,046	86,468
R (The calculations without the seal structure)	66,123	86,715
A (The experimental results)	68,000	84,000

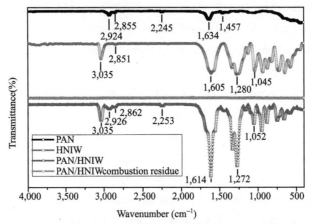

Figure. 12.1 FT-IR spectra of fibrous film, HNIW, composite of PAN and HNIW, and combustion residue of composite (with 70 wt% HNIW).

2) Clear organization

One of the suggestions to organize a paper for a successful oral presentation is to proceed in the same logical pathway that you do in writing a paper and structure the ideas in an easy-to-follow way. For doing so, think about the following two questions:

(1) Do you put what you will say in a logical sequence?

(2) Do you write down key words and expressions to remind you of the main points?

In order to answer the first question, you have to know what may be "a logical sequence." Basically, the answers are various: it can be the way you write the paper, the process of conducting the experiment, the cause-and-effect explanation of a phenomenon and so on. The logical sequence should be shown at both macro-level and at micro-level. The former refers to the organization of the entire presentation while the latter is the organization of one or several inner related slides. Logical sequence at macro-level is always shown in the organization of ideas throughout the slides: title slide→ purpose slide→ contents slide→ body slides→ conclusion slide→ references slide (Cong and Li, 2007:128).

E. g. 5:

Contents
1. Introduction
2. Materials and Methods
➢ Preparation
➢ Characterization
➢ Photo catalytic degradation
3. Results and Discussion
4. Conclusions
5. References

(From a student's homework)

It is the second slide in the presentation that immediately tells the audience the five main parts in the paper:

The **introduction** section: to introduce the research scope, what has been done, what has not been done and the primary purposes of the paper.

The **materials and methods** section: to explain how to prepare and conduct the research.

The **results and discussion** section: to reveal the major findings and show the significance of the research by analyzing the relationship among the observed facts.

The **conclusion** section: to summarize the research and suggest the future research lines.

The **references** section: to list the works cited.

Second, keep slides simple for the audience to grasp the main points immediately. The slides should just contain the most necessary information. To fulfill this purpose, you should write clearly and concisely in **bullet points**. A **bullet point** is a thing in a list that consists of a word or short phrase, with a small printed symbol in front of it (Longman Dictionary of Contemporary English, 2004). It is usually marked by a square or round symbol. Bullet points can highlight important points, simplify ideas and enhance understanding. The following example is an effective one.

E. g. 6:

Large Scale Image Retrieval

What are the applications of image retrieval?
- *Search engine;*
- *News reports;*
- *Online shopping;*
- *Others.*

(From a student's homework)

Study the following slide, is it an effective one? If not, what is the problem with it?

E. g. 7:

---

a) convergence-straight duct

When the sub-munitions fly at high speeds: the inlet airflow velocity is high, the exit airflow velocity changes randomly with the sub-munitions' flying speed, also is the turbine rotation speed, that calls for higher mechanical strength of all mechanical parts especially the turbine and shaft section. When the sub-munitions fly at a low speed: the generator may lack of sufficient rotation speed. So convergence-straight duct unsuitability applies on the sub-munitions fuses.

(From a student's homework)

This one is confusing because it contains overcrowded information on one single slide and the lettering is so small that the audience cannot see clearly the main points in a few seconds. They may feel frustrated and lose interest in your presentation. So you have to summarize just a few key words and expressions to permit a rapid reading, not always paragraphs of complete sentences. You can write bullet points to highlight the structure and the main ideas of the presentation. All in all, each slide should look simple, and be easy to be understood at the time the slide is on the screen; it should not repeat what you are saying. Remember both the presenter and the audience are not PPT readers, what you can do is to help them grasp the key points instead of failing them.

Last but not least, size font, typeface and brightness of color are also important, which should be adjusted to be in accordance with the contents and its organization.

3) Efficient expression

Before a presenter comes to the stage with well-structured manuscript and carefully-designed PPT slides, he has to take these questions into consideration: How can I make the communication more successful? How can I maximize the effect of my written language? Here are three considerations for efficient expression of presentations:

Appearance;

Vocal delivery;

Rapport with the audience.

(1) Appearance.

Appearance does not simply mean the clean and neat dress but also your facial expression. You look right and you speak right. While, what can be seen as "the right expression on the face"?

That is **CCRPPE** principle: being confident, calm, relaxed, pleasant, proud and enthusiastic. These words are easy to understand but may not be easy to stick to. The following tips can also be helpful:

a. Walk slowly and confidently.

b. Keep good postures: Stand upright without frequent body movement.

c. Lower your shoulders and hide physical signs of anxiety.

d. Keep a smile on your face.

e. Be proud of what you are talking about, but not arrogant.

f. Wish to share your ideas with others.

g. Look serious but not over-nervous.

Remember the above tips and apply them to your rehearsals and presentations, practice with your visual materials and find out the time for each slide to know how long you will discuss it with the audience (in general the time with one slide is about one minute or less).

(2) Vocal delivery.

Voice control is another key factor. To have yourself heard and understood by the audience you have to focus on two things: audibility and pace/speed variations. To make the back-row audience hear you clearly, speak very clearly and project your voice. Do not lower your head and mumble, which will make the audience feel frustrated. Do not shout, which will make the audience too excited to concentrate on the speech. If you are not a native speaker especially when facing a large group of audience, avoid speaking fast: the larger the audience is, the slower the delivery should be.

More importantly, a speaker must not always speak slowly. He has to change his speed and show the speed variations here and there: When it comes to the main points, speak slowly with more strength; when it comes to the illustration of supporting details, speak faster with fewer pauses. Read the following example and find out which words the American president Obama speaks slowly.

E. g. 8:

Good evening. Tonight, I can report to the American people and to the world that US has conducted an operation that killed Osama Bin Laden, the leader of Qaeda, a terrorist who is responsible for the murder of thousands of innocent men, women and children.

Here **tonight, American people, killed, Osama Bin Laden, murder, thousands, men, women** and **children** are important information about **when, who, what** and **why**, at which the president slows down his speed with pauses and stress to let the audience be clear about the event.

(3) **Rapport with audience**（与听众良好的互动）.

Establishing good relationship with audience and knowing their feedback can improve their understanding of important points and help you assess how your presentation is received. Although in the process of presentation, the presenter usually holds the floor and is not expected to be interrupted by the audience, he or she can communicate with the audience via eye contact and body languages.

For example, you can give individual attention to their feedback, such as direct eye contact to the audience; you can also use some gestures to strengthen the understanding in explaining main points.

E. g. 9:

Read the following sentences from a speech by Professor Rod Ellis about how to provide the non-native students with corrective feedback in their writing and tell at which words and expressions hand gestures are used and how.

Should the teachers correct all the single error the student has made? Or they deliberately focus on article errors or past tense errors and then change from one time they focus on one type of error, the next time focus on another type of error, etc.

In fact, professor Rod Ellis uses appropriate hand gestures (he puts his hands first on the left, then on the right to show the change of the different error types) at these following expressions to maximize the effect of his spoken language since these expressions are key to the understanding of his speech:

article errors	**past tense errors**
one time one type	**next time another type**

Besides hand gestures, there are more suggestions for communication with the audience, the following five may be helpful:

a. Have direct eye contact with a number of people in the audience.

b. Every now and then, glance at the whole audience while speaking.

c. Better for you to stand, walk or move about with appropriate hand gesture.

d. If you make an error, correct it, and continue.

e. Remember to thank your audience at the end of the presentation.

12.3 The Question-and-answer Session

The Question-and-answer session (shortened as Q&A session) is always included as a part of paper presentation at international conferences. The time allotment is about 15 minutes to 30 minutes. After your presentation, the chairman declares the Q&A session, gets the questioning started by some remarks to break the possible silence of tension and invites questions. Obviously a successful Q&A session has many advantages: The message advanced by the speaker is reinforced through the recall and collaboration of its important points (Cong and Li, 2007:138); it is a good chance for the audience to hunt for more information they are interested in; the speaker can collect valuable feedback information. However it is not so easy to handle the questions for they could be extensive, unpredictable even irrelevant. Moreover, there is no time for preparation or delay, the speaker usually bears high pressure.

1) Purposes of questions

In most cases the audience ask questions because they have unclear points, misunderstandings or problems in the presentation. In other cases the audience just want to know more details about your research and how your research may relate to their researches. Generally, there are 8 purposes of question-raising as the following.

a. Clarifying problems.

b. Showing special interests.

c. Raising different opinions.

d. Hunting for information.

e. Conducting comprehensive examinations.

f. Other reasons (hinting plagiarizing, challenging, attracting other's attention, etc.).

2) How to answer questions

Most of the questions in the Q&A session are ordinary and easy to answer. Take 3 steps in answering them. First, repeat the question. It may not be a word to word repetition, but a paraphrase of the original one. To do so you not only explain the question to other audience but also analyze it and think about it quickly. Second, provide your answer to the question and keep it to the point. Third, give reassurance by offering a direct reference to the question-raiser to see if the explanation is enough or further information is needed.

For the questions you lack an answer, do not panic or bluff, admit that you do not know (Day and Gastel, 2007:172). Be honest and do not worry about that you will lose respect of the audience. If you know how to find the answer, offer the information sources. To avoid this kind of trouble, you should invite your colleagues to ask you some would-be questions when you rehearse, and the more the better.

For questions of disagreement, respect the facts, have sincere attitude and persuade the question-raiser with evidence. Do not be stubborn and put yourself into embarrassment.

For noncommittal questions, such as the questions which are sensitive, inappropriate to answer, have diplomatic attitude. Answer questions carefully, and offer reserved and flexible answers. If you feel it is really troublesome and time-taking to answer, shift to another topic or imply a further discussion after the Q&A session.

3) Tips for Q&A session

To make your Q&A session successful, you had better remember the following five tips.

a. Prepare some questions in your rehearsal.

b. Offer direct answer and do not turn it into a long story.

c. Be objective. Explain your point of view with evidence. Do not be arrogant.

d. Answer questions with room for improvement.

e. Be polite and use subjunctive mood.

12.4 Summary

Presentations play an important role in academic communication. Good presentations can not only spread your new findings but also enhance the understanding among professionals. To make your presentation successful, first know the three features: good preparation, clear organization and efficient expression. Then remember how to do so and rehearse with slides. Anyway, you will be more skillful through hard practice.

Tasks

Task 1

Directions: Prepare and make a 5-minute presentation of your current research interest to your peer. Then vise versa. Take notes of your peer's presentation and give feedback.

Task 2

Directions: Work with your peer and revise the following slide.

Radar

can be used for
- military defense;
- report weather;
- navigate and locate.

can be used to
- monitoring natural disasters;
- monitoring agricultural crop;
- survey the store of mineral resources.

Chapter 13
How to Create Scientific Posters

Nowadays, poster sessions have become a common feature at scientific conferences. A scientific poster is a visual presentation of research work by an individual or research teams in an interactive (face-to-face) and social setting at a congress or conference with an academic or professional focus (Allard, 2012). When presented well, posters can be more effective than a talk in establishing a relationship with your audiences since such presentations allow you to interact one on one with the people who are interested in your research. In a poster session, people usually walk around with food or drinks to search what may catch their attention with various sizes according to different conferences attached to the wall and researchers need to stand in front of their posters and wait for communicating with people who are interested in the topics or contents.

Generally, posters have a very different orientation from written papers but still follow the standard scientific format in research papers except the Discussion. Posters, therefore, contain: Title, Introduction, Materials and Methods, Results, Conclusion, and Acknowledgments. Optional are Abstract and References (Hofmann, 2014: 522). This chapter aims to improve the quality of posters and to provide guidelines to those who might be approaching their first professional conference.

13.1 Basic Components

Creating posters is distinguished from writing a manuscript. Poster sections concentrate only on the main points and present these sections briefly and visually (Hofmann, 2014). Otherwise, a poster will suffer if it includes too much information. Every component, therefore, should indicate key points rather than tedious information. The following sections will be introduced separately.

1) Title

A title is undoubtedly one of the most important parts of posters. The purpose of a title is to attract busy viewers during posters session, so that they will want to access and read your poster. The more revealing your title is, the more easily your potential viewers can judge how relevant your poster is to their interests. Here are two pieces of advice.

(1) Use key words preferentially.

A title should be very effective, brief, informative, and with key words placed near the front. Thus it is important to decide which words (keywords) will capture the attention of viewers likely to be interested in your poster and to place them near the front of your title (Cargill and O'Connor,

2013). In these conditions, titles should avoid beginning with vague terms such as "the research...," and "The effects...". See Example 1:

1a. *"Effects of substance A on protein B"*;

1b. *"Substance A inhibits protein B."*

Title 1b is result-oriented, thus obviously is more efficient than title 1a. Using a title like example 1b can provide more information, so that viewers at least have something to take away in terms of message from your work even if they don't read contents of your poster. The other advice shows below.

(2) Keep concise.

A title should convey the topic, the approach, and the organism but keep concise at the same time. Reduce any redundant information down to at most two lines. That is to say, do not make your title spill onto three lines.

Another point is that the title is usually combined with a logo of your working institution. See Example 2:

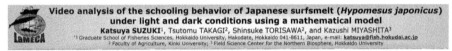

(Cited from Suzuki's poster, 2004)

2) Introduction

Your reader was mildly intrigued by the title, but you have exactly two sentences to hook them into reading more. In Introduction, according to applied linguists, there are usually five stages. Firstly, state about the field of research with a setting or context for the problem to be investigated. Secondly, present more specific statements about the problem already studied by other researchers than laying a foundation of well-known information. Thirdly, indicate the need for more investigation and create a gap or research niche for the present study to fill. Fourthly, give purpose or objectives of the writer's study or outline its main activities or findings. Finally, if possible, give a positive value or justification for carrying out the study (Hofmann, 2014: 528). However, there is no room for all stages of an introduction in a poster. The rule you need to follow is "do not overwhelm your readers with background information." Provide only the minimum background information needed to capture their attention. And then describe exactly what your interesting question is and why it is really needed to be addressed. In addition, provide a brief description and justification of the general experimental approach. It is better to keep it within 150 words (Example 3) and if you want to create a more attractive Introduction, you could also make a combination of pictures or figures with words (Example 4).

Example 3:

INTRODUCTION

While thermal and tactile stimulations have been investigated using functional neuroimaging, a number of questions still exist regarding these basic perceptual processes. For example, studies investigating thermal changes have primarily examined the stimulation applied to only one hand, yielding no knowledge related to the differences in the laterality of thermoception (Craig, 1995; Hua, Strigo, 2005). This is a fundamental question, as other studies have identified laterality differences in tactile stimuli (Coghill, Gilron, 2001). Additionally, thermal and tactile changes have never been directly compared, resulting in little knowledge of how these systems are similar or different. In this poster, we describe a new apparatus designed to apply parametrically-varied thermal and tactile stimulations to each hand independently. This apparatus was constructed with the intention of building event-related fMRI experiments to examine both independent and combined thermal and tactile stimuli.

(Cited from Bennett's poster, 2010)

Example 4:

Introduction

- Honest signals are traits that individuals within a population possess that reflect their quality
- Honest signals are often visual characteristics (plumage size and color, bill color, foot color, etc.)
- Honest signals should provide a potential mate with information about the physiological and/or genetic quality of another individual
- Assortative mating is a process of nonrandom mating in which mates choose each other based on similarities in phenotypes

Adult Black Guillemots
Photo Credit:
http://sciences.umf.maine.edu/images/guillies.jpg

- The black guillemot is a monomorphic seabird characterized by its black and white body and strikingly red feet
- Foot color has been shown to affect mating behavior in another monomorphic seabird, the blue-footed booby (*Sula nebouxii*) [1,2]
- Recently, foot color in guillemots has been linked to oxidative stress levels, suggesting that red feet may be honest signals of condition in these birds [3]

(Cited from Carlton's poster, 2007)

3) Material and Methods

Traditionally, Material and Methods section provides the necessary information for another competent scientist to repeat the work and establish credibility for the result, thus you need to give enough details about how the work was done (Cargill and O'Connor, 2013). However, it is not the case in such section of a poster. In this section, you should summarize your experimental approach very briefly. Use illustrations such as flowcharts, photographs, drawings or schematics to present your experimental procedures rather than words. Researches have shown that verbosity would drive readers away in a poster session. Example 5 and Example 6 showed the opposite expression of Material and Methods section.

Example 5:

MATERIAL and METHODS

Subjects. Investigators conducted this study in outpatient clinics of six family medicine residency programs across Texas. Patients were invited to participate if they were adults with low back pain for 3 months or longer, and were not new to the clinic. Investigators excluded pregnant women and patients with cancer.

Procedure. Medical students enrolled and surveyed 222 patients as they arrived for routine visits to the outpatient family medicine clinics. When the visit was complete, students retrieved their medical records and abstracted additional information related to low back pain.

Measurement. The 5-page patient survey addressed demographic characteristics, pain duration, frequency and severity, physical functioning and general health, anxiety, depression, social support and stress, and family violence. From the charts, students gathered information about the duration of the doctor-patient relationship, the patient's health insurance, the cause and duration of the low back pain, treatments for pain, comorbidities, and BMI.

Analysis. In this analysis, the key outcome variables were average pain and total body pain. Total body pain is measured by combining average pain with its effect on normal daily work.[4] Predictor variables included BMI as well as diagnoses including diabetes, hypertension, and hyperlipidemia.

(Cited from Lai's poster, 2008)

Obviously, the presenter in Example 5 explains how the study was conducted literally, while Example 6 uses words combined with photographs and a diagram to show experimental procedures, which provides very brief and clear illustrations.

Example 6:

METHODS

The design of the apparatus involved the use of four Peltier-effect thermoelectric devices for thermal stimulation and ten pneumatically-driven wooden pegs for tactile stimulation. These elements were embedded within wood blocks designed to fit the contours of an average human hand (see below). Each hand block (one each left and right) contained two thermoelectric devices, five tactile stimulators, and one thumb button for responses. To enable the rapid application of large thermal changes each thermoelectric device was cooled with water pumped from the control room. Thermal and tactile stimulus application was controlled through the use of a programmable microcontroller. This control system enabled reliable millisecond timing for event-related stimulus application.

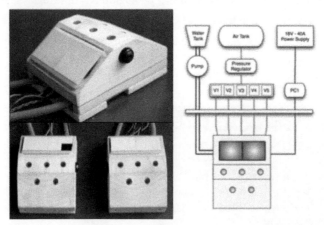

(Cited from Bennett's poster, 2010)

4) Results

Results section in a poster shares the same objective and functions with the section in a traditional paper. It functions in highlighting the important findings, locating the figure(s) or table(s) where the results can be found and giving comments on the results. For basic scientific posters, this section is the largest portion of the poster. For the sake of conciseness, however, it should briefly describe qualitative (定性的) and quantitative (定量的) results using bullets (项目符号或编号) or limited text (有限的文字). Most of your findings should be presented in tables and figures. Use figures with figure legends (图表说明), graphs and tables to enhance the presentation of your results, which can stand on your own and interpret your findings, not just a textual description. Readers are far more likely to stop at a poster if there are colorful, high quality images, such as Example 7.

5) Conclusions

For the Conclusions section, assign importance to your results and summarize them accordingly. You need concentrate on your main findings and their interpretation rather than a list of all your research findings. Do not forget to relate your findings to other published researches in the field. This movement would give your research more impact on your readers and show your professionalism at the same time. In addition, tell your readers why your research is interesting or significant. Use the same key terms you used consistently throughout the poster. Moreover, it is better to make the use of bullets, numbered lists, arrows (箭头), italics, or colored text to emphasize major points. Keep it brief and within 200 words. Please keep it in your mind that the Conclusions section is usually the first part conference audience will start reading besides the Title, which means you really need to hook them to read more about your poster. An example of a well-designed Conclusions section is shown in Example 8.

Example 7：

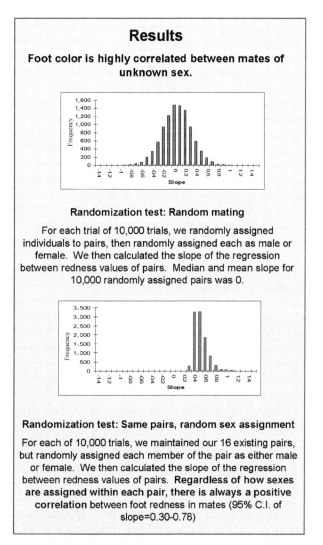

(Cited from Carlton's poster, 2007)

Example 8：

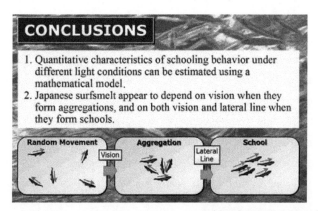

(Cited from Suzuki and Takagi's poster, 2004)

In Example 8, the writer uses numbered lists and flowcharts (流程图) to present the research findings, which makes this section very concise and easy to understand. Here the supplementary is that there are other types of combination of sections shown in posters. For instance, the Results and Conclusions are displayed on the same panel at times. Sometimes, the Conclusions section is followed by a brief section called "Further Research" in which you explain briefly how you are planning to extend the present research.

6) References and Acknowledgments

If you cite information within the poster text, references should be included. However, keep references to a minimum (no more than five) and keep them brief. In the interest of acknowledgments, if applicable, thank individuals for specific contributions (equipment donation, statistical advice, laboratory assistance, comments on earlier versions of the poster); mention who has provided funding; be sincere but do not release too much into informality in this section and do not list people's titles.

13.2 Format

The format of a poster is so vital that it can determine whether the poster is effective. In this part, we will provide information about how to tackle format details of a poster, namely, Layout, Headings, Text, and Colors.

1) Layout

Unlike a traditional research article, a poster can adopt various layouts. However, viewers expect to find certain sections they are familiar with in specific places. For instance, the most important text sections such as Introduction or Conclusions are usually placed in the top left and bottom right. The most important visuals should be placed in the middle of a poster. The least important sections and visuals are often placed on the very bottom. One of the first things you need to figure out is that how much space you are allowed to use. Generally, presenters are allotted a space as 1 m × 1.3 m or as large as 1.3 m × 2 m. You can, usually, make your poster in these two sizes: A1 (841 mm × 594 mm) and A0 (85 cm × 120 cm). The size of a poster you use depends on the requirement of the conference you attend. Here are samples of poster layout in Example 9 and Example 10.

Here are seven pieces of advice from other experts:

(1) Use a visual grammar to guide readers to the important parts of your poster.

(2) For maximum visual layout and impact, aim for about 20% text, 40% graphics and 40% empty space (Hofmann, 2014).

(3) Use different sizes and arrangements for the various poster sections.

(4) Use headings intelligently to help readers find your main points and key information. Balance the placement of text and graphics to create visual appeal.

(5) Use a column format to make your poster easier to read in a crowd.

Example 9　The layout of size A0:

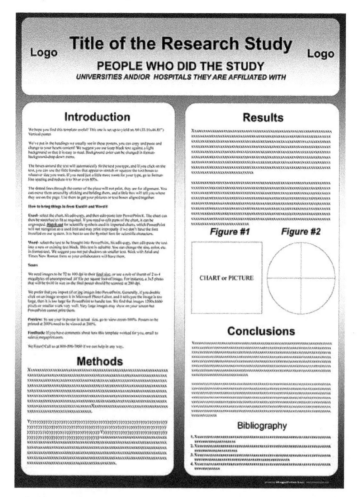

Example 10　The layout of size A1:

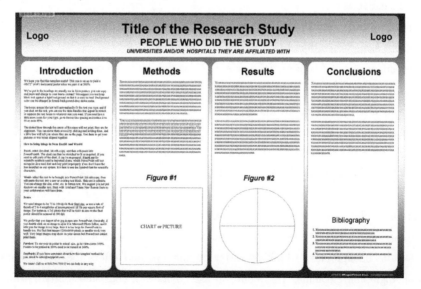

(6) Use "reading gravity" (阅读重力) which pulls the eyes from top to bottom and left to right (Wheildon, 1995) to design your poster's layout.

(7) Use white space creatively to help define the flow of information (source from colinpurrington. com).

2) Headings

Headings as an organizational construction orient readers, convey key points and help viewers find what they are looking for. The Headings here include Titles, sub-titles, section titles, and figure legends.

There are some rules you need to follow when you are dealing with the Headings. Firstly, use headings as opportunities to summarize your work in large letters and make it as a flow throughout the poster so that it can guide your readers. A hurried reader should be able to get the main points from the headings alone. Good headings are part of the visual grammar that help move readers through your poster. Let your headings be hierarchical, which means the more important the point, the larger the type. Finally, make the strong statements bold, and the font size of each heading should be large enough to be read from 1.5 meters away.

3) Text

The main objective in preparing text for a poster presentation is to edit it down to very concise language. Know that people are attracted to posters that have good graphics, a clear title, and few words. It is no doubt that an effective poster is corresponding with text format (fonts and sizes). Suggestions are shown below.

(1) Minimize text — use images and graphs instead.

(2) Font sizes should be large enough to be read from 1.5 meter away.

(3) Use a sans-serif font (无衬线字体,匀称字体) (e. g. Arial) for most text, which is easier to read than serif fonts (衬线字体,非均匀字体) (e. g. Times New Roman).

(4) Use phrases rather than full sentences.

(5) Use an active voice.

(6) Left-justify text; avoid centering and right-justifying text.

(7) Use font sizes proportional to importance (Hofmann, 2014).

(8) Use sentence case (仅第一个字母大写) for the Title; lowercase lettering (小写字母) for text with initial capitals where needed.

(9) Pay attention to text size in figures — it must also be large.

(10) Use italics (斜体) or bold (粗体) to add emphasis.

4) Colors

The color design is important in the presentation of your data. Therefore, you need to be very cautious when using colors. Use a colored background to unify your poster and you may also use a second background color to frame individual sections of the poster, or use colored text boxes on a white or light colored background to add visual interest. For the sake of hooking your readers, use muted color for the background, which is easier on the eye and offers the best contrast for text, graphic and photographic elements. It is better to use colors in a consistent pattern; otherwise, your poster will be distractive rather than attractive. If you want to make your poster colorful, then stick to a theme of two or three colors since much more will overload and confuse readers.

13.3 Software

It would be more professional to generate a poster electronically by using a variety of software packages such as Microsoft PowerPoint, Adobe Photoshop, Keynote, and Pages. We will provide only an illustration about how to use PowerPoint to generate a poster due to time and space limitations.

First, open a new file and choose the blank page as your layout. To design a large poster, you must tell PowerPoint (or whichever program you're using) how large the paper is. You can do this by going to the **File/Page Setup** menu — just enter the width and height you want, within the limitations given above. It will choose Portrait or Landscape automatically, based on the height and width you enter. Once you have done this, press **OK**. PowerPoint may complain that the size exceeds that of the current printer, but don't worry, just press **OK** to continue. You should now be facing a blank page in the appropriate dimensions. If the rulers are turned on, you will see that it's the size you asked for. Then you can start to design your poster by choosing background colors, text format and so on.

Moreover, there is a simple and quick way, that is, using the poster templates on the website www.posterpresentation.com. This website provides lots of poster templates in different sizes that can meet various requirements.

13.4 Presenting

When all these things about designing and printing a poster get done, there comes the poster presentation, a crucial step. If you cannot present your poster well which makes all readers remember you and your work, you will waste your designing.

For an effective poster presentation, you should arrive early at the display site, and bring a poster hanging kit (布置海报的工具包) with you unless you are confident the conference organizers will have proper supplies. It is important to hang your poster squarely (垂直地) and neatly (牢固地) and make sure you are at your poster spot during your assigned presentation time. Do not forget to bring copies of a handout for your readers, which should include a miniature version of your poster and more detailed information about your work, in an illustrated narrative form. For better involvement with readers, consider leaving a pen and pad inviting comments from viewers. Use your poster as visual aids during your presentation, but do not read your poster. What you should focus on while you are presenting is to tell viewers the context of your problem and why it is important, your objective and what you did, what you discovered, and what the answer means in terms of the context.

13.5 Summary

This chapter has provided essential guidelines about how to prepare, create and present a scientific poster, including basic components, format, software and presenting keys. If you earned the opportunity to attend an international conference for your first time, you are encouraged to offer a poster presentation rather than an oral presentation. This is reasonable advice, because when presented well, posters can be more effective than a talk in establishing a relationship with your audience and may help you become familiar with academic routines efficiently.

References

Andrew, P. On the Limiting Behaviour of Lévy Processes at Zero Probab[J]. *Theory Relat. Fields*, 2008,140:103-127.

Allard, A. W. Creating Poster Presentation. In Beins, B. C. and Beins. B. M. *ed*. Effective Writing in Psychology: Papers, Posters and Presentations [M]. John Wiley & Sons, Inc. , 2012.

Anthony, L. Characteristic Features of Research Article Titles in Computer Science [J]. *IEEE Transactions of Professional Communication*, 2001, 44(3): 187-194.

Ayers, G. The Evolutionary Nature of Genre: An Investigation of the Short Texts Accompanying Research Articles in the Scientific Journal *Nature*[J]. *English for Specific Purposes*, 2008, 27 : 22-41.

Baker, P. How to Write Your First Paper [J]. *Obstetrics, Gynaecology and Reproductive Medicine*, 2011, 22:3.

Bennett, C. M. How Reliable Are the Result from FMRL? [poster]. Cognitive Neuroscience Society Meeting, 2010.

Bennett, C. M. A Device for the Parametric Application of Thermal and Tactile Stimulation During FMRL [C]. Organization for Human Brain Mapping Meeting, 2010.

Bhatia, V. K. Analyzing Genre: Language Use in Professional Settings [M]. London: Longman Press, 1993.

Bizup, J. and Williams, J. Style: Lessons in Clarity and Grace (11th edition) [M]. Essex: Pearson Education, 2014.

Blanch, J. ; Lenehan, E. and Quinton, S. Dispersant Effects in the Selective Reaction of Aryl Diazonium Salts with Single-Walled Carbon Nanotubes in Aqueous Solution [J]. *J. Phys. Chem. C*, 2012, 116: 1709 – 1723.

Blouin, N. ; *et. al*. Toward a Rational Design of Poly(2,7-carbazole) Derivatives for Solar Cells [J]. *Journal of the American Chemistry Society*, 2008,130:732-742.

Brittman, F. The Most Common Habits from more than 200 English Papers Written by Graduate Chinese Engineering Students [J]. *Science Net*, 2009.

Camp, R. J. ;*et. al*. Molecular Mechanochemistry: Low Force Switch Slows Enzymatic Cleavage of Human Type I Collagen Monomer [J]. *Journal of the American Chemical Society*, USA & UK, 2011(133): 4073-4078.

Candan, K. S. ; Rangan, P. V. and Subrahmanian, V. S. Collaborative Multimedia Systems: Synthesis of Media Objects [J]. *IEEE Transactions on Knowledge and Database Engineering*, 1998, 3 (10): 433-457.

Cargill, M. and O'Connor,P. Writing Scientific Research Articles: Strategy and Steps (second edition) [M]. Oxford: Wiley-Blackwell, 2013.

Carlton, E. D. Assortative Mating by Foot Color in the Black Guillemot (poster). www. biology. kenyon. edu, 2007.

Carter, S. ; Hunt, B. and Rimmer, S. Highly Branched Poly(*N*-isopropylacrylamide)s with Imidazole End Groups Prepared by Radical Polymerization in the Presence of a Styryl Monomer Containing a Dithioester Group [J]. *Macromolecules*, 2005, 38: 4595-4603.

Cong, C. and Li, Y. Y. English for Communication [M]. Nanjing: Press of Nanjing University,2002.

Cover letter. http://emuch. net/html/201101/2787763. html.

Dass, S. C. Assessing Fingerprint Individuality in Presence of Noisy Minutiae [J]. *IEEE Transactions on Information Forensics and Security*, Vol. 5, No. 1, March,2010.

Davis, B. L. and Wang, M. Y. Technical English Writing for Graduate Students and Professionals [M]. Beijing: Higher Education Press, 2006.

Day, R. A. How to Write and Publish a Scientific Paper [M]. Cambridge: Cambridge University Press, 1988.

Day, R. and Gastel B. How to Write and Publish a Scientific Paper [M]. Beijing: Peking University Press, 2007.

Ellis, M. and Christofides, P. D. On Finite-time and Infinite-time Cost Improvement of Economic Model Predictive Control for Nonlinear Systems [J]. *Automatica*, 2014 (50): 2561-2569.

Emerald Group Publishing. How to Survive Peer Review and Revise Your Paper [EB/OL]. URL: http://www.emeraldgroup publishing.com/authors/guides/promote/review.htm (retrieved on 20th May, 2015)

Fayolle, A. and Wright, M. (eds). How to Get Published in the Best Entrepreneurship Journals: a Guide to Steer Your Academic Career [M]. Cheltenham: Edward Elgar, 2014.

Flowerdew J. and Mahlberg M. Lexical Cohesion and Corpus Linguistics [M]. Philadelphia John Benjamins Publishing Co., 2009.

Fontanet, I.; Coll, J. F.; Palmer, J. C. and Posteguillo, S. The Writing of Titles in Academic Research Articles [C]. In R. M. Chamorro, A. R. Navette (eds). *Lenguas Aplicadas a las Cienciasy la Tecnologia Approximaciones*. Caceres, Spain: Universidad de Extremadura, Servicio de Publicaciones, 1997: 155-158.

Longman Dictionary of Contemporary English, 4th edition [M]. BeiJing: Foreign Language Research and Teaching Press, 2004.

Furfaro, R.; et. al. Asteroid Precision Landing via Multiple Sliding Surfaces Guidance Techniques [J]. *Journal of Guidance, Control, and Dynamics*, Vol. 36, No. 4, July - August, 2013.

Glasman-Deal, H. *Science Research Writing for Non-Native Speakers of English* [M]. Beijing: China Machine Press, 2011.

Gopen, G. and Swan, J. Science of Scientific Writing [J]. *American Scientist*, 1990 (78): 6.

Halliday, M. A. K. and Hasan, R. Cohesion in English [M]. Beijing: Foreign Language Teaching and Research Press, 2001.

Harris, M. W. and Açikmeşe, B. Lossless Convexification of Non-convex Optimal Control Problems for State Constrained Linear Systems [J]. *Automatica*, 2014 (50): 2304-2311.

Hartley, J. To Attract or to Inform: What Are titles For? [J]. *Journal of Technical Writing and Communication*, 2005, 35(2): 203-213.

Hays, J. C. Eight Recommendations for Writing Titles of Scientific Manuscrips [J]. *Public Health Nursing*, 2010, 27 (2): 101-103.

Hofmann, A. H. Scientific Writing and Communication—Papers, Proposals, and Presentations [M].2nd. Oxford: Oxford University Press, 2014.

Hyland, K. Disciplinary Discourses: Social Interactions in Academic Writing [M]. London: Longman, 2000.

Hyland, K. Metadiscourse [M]. Beijing: Foreign Language Teaching and Research Press, 2008.

Hyland, K. Academic Discourse [M]. Continuum International Publishing Group, 2009.

How to Submit a Revision and Tips on Being a Good Peer-Reviewer [EB/OL]. http://www.ncbi.nlm.nih.gov/pmc/articles/PMC3970178/

Hu, G. S.; Li, P. F. and Shen, Y. Z. English Paper Writing and Publication [M]. Beijing: Higher Education Press, 2010.

Huang, X. English Scientific Writing Course [M]. Beijing: Chemical Industry Press, 2009.

Joqulekar, A. S.; Damodaran, K. and Kriel, F. H. Dictyostatin Flexibility Bridges Conformations in Solution and in the β-Tubulin Taxane Binding Site [J]. *Journal of the American Chemical Society*, 2011, 133 (8): 2427-2436.

Lai, J. Common Medical Diagnoses Associated with Chronic Low Back Pain [EB/OL]. http://

familymed. uthscsa. edu/rrnet/students. asp, 2008.

Langan, J. College Writing Skills with Readings (Sixth Edition) [M]. Beijing: Foreign Language Teaching and Research Press, 2007.

Li, P. and Cao, Y. Abstract Noun Structure in English Titles of Scientific Articles [J]. *Chinese Journal of Scientific and Technical Periodicals*, 2012, 23 (2): 322-324.

Livnat, Z. Dialogue, Science and Academic Writing [M]. John Benjamins Publishing House, 2012.

Mabe, M. A. and Amin, M. *Dr.* Jekyll and Dr. Hyde. Author-reader Asymmetries in Scholarly Publishing Aslib Proceedings, 2002, 54 (3): 149-157.

Ma, W. Y. Comparative Analysis Word Usage in Titles Between Chinese and American High Impact Chemical Journals [J]. *Chinese Journal of Scientific and Technical Periodicals*, 2014, 25 (1): 65-68.

Manuscript format [EB/OL]. http://en. wikepedia. org/wiki/Manuscript format.

Nadkarni, M. ; Ohno-Machado and Chapman, W. Natural Language Processing: an Introduction [J]. *J Am Med Inform Assoc.*, 18(2011):544-551.

Neff, J. G. and Prues, D. (*et. al.*) Formatting and Submitting Your Manuscript [M]. Ohio:Cincinnati, 2000.

Peer-review Policy [EB/OL]. http://www. nature. com/authors/policies/peer_review. html.

Peh, W. C. G. and Ng, K. H. Preparing a Manuscript for Submission [J]. *Singapore Medical Journal*, 2009, 50 (8): 759-762.

Powell, K. Publications: Publish Like a Pro[J]. *Nature*, 2010, 467: 873-875.

Richert, D. and Cortés, J. Optimal Leader Allocation in UAV Formation Pairs Ensuring Cooperation[J]. *Automatica*, 2013, 49: 3189-3198.

Rosen, L. J. ; Behrens, L. and Branscomb, H. E. Allyn & Bacon Handbook [M]. NJ: Prentice Hall College Div. , 1999.

Sanjeev, B. ; Sumedh, W. S. ; Kishore, N. M. and Thomas, M. C. MPS: Miss Path Scheduling for Multiple-issue Processors [J]. *IEEE Transactions on Computers*, 1998, 12 (47): 1382-1397.

Socolofsky, S. A. How to Write a Research Journal Article in Engineering and Science [EB/OL]. Dept. Civil Engrg. , Ocean Engrg. Div. , Texas A&M Univ. , M. S. 3136, College Station, TX 77843-3136. URL: http://www. ifh. uni-karlsruhe. de/lehre/dokkurs/ gutes-schreiben/paper_how-to. pdf 2004. (retrieved on 20th May, 2015)

Soler, V. Writing Titles in Science: an Exploratory Study [J]. *English for Specific Purpose*, 2007, 26: 90-102.

Srivastava, C. ; *et. al.* Arabidopsis Plants Harbouring a Mutation in AtSUC2, Encoding the Predominant Sucrose/Proton Symporter Necessary for Efficient Phloem Transport, Are Able to Complete Their Life Cycle and Produce Viable Seed [J]. *Annals of Botany*, 2009, 104: 1121-1128.

Suzuki, K. Video Analysis of the Schooling Behavior of Japanese Surfsmelt Under Light and Dark Conditions Using a Mathematical Model (poster). The North Pacific Marine Science Organization Thirteenth Annurral Meeting, 2004.

Swales, J. M. Genre Analysis: English in Academic and Research Settings [M]. Cambridge University Press, 1990.

Swales, J. M. and Feak, C. B. Academic Writing for Graduate Students[M]. Michigan: The University of Michigan Press, 2010.

Tahat. Aluminium-silicon Alloy and Its Composites Reinforced by Silicon Carbide Particles [J]. *Microelectronics International*, 2010, 27(1): 21-24.

Tarone, E. ; *et. al.* On the Use of the Passive and Active Voice in Astrophysics Journal Papers: With Extensions to Other Languages and Other Fields [J]. *English for Specific Purposes*, 1998, 17(1): 113-132.

Tsujiuchi, N.; et. al. Modeling and Control of a Joint Driven by Pneumatic Actuator [J]. *IEEE*, 2009, 2272-2276.

Van Rys, J.; Meyer, V. and Sebranek, P. The Research Writer: Curiosity, Discovery, Dialogue [C]. Belmont CA: Wadsworth Publishing Company, 2011.

Wallwork, A. English for Writing Research Papers [M]. New York: Springer, 2011.

Wang, L. Y. and Wang, L. Z. Textual Analysis of English Titles in Chinese and America Medical Journal: A Study Based on Borpus Linguistics [J]. *Chinese Journal of Scientific and Technical Periodicals*, 2011, 22(5): 784-788.

Wescott, B. L.; Stewart, S. D. and Davis, W. C. Equation of State and Reaction for Condensed-phase Explosives [J]. *Journal of Applied Physics*, 2005, 98: 053514.

Wheildon, C. Type and Layout [M]. Berkeley: Strathmoor Press, 1995.

Williams, P. S.; et. al. Coupling of Estimation and Sensor Tasking Applied to Satellite Tracking [J]. *Journal of Guidance, Control and Dynamics*, Vol. 36, No. 4, July-August, 2013.

Winkler, A. C. and McCuen-Metherell, J. R. Writing the Research Paper (*a Handbook, Seventh Edition*) [M]. Beijing: Peking University Press, 2008.

Wolfe-Simon, F.; et. al. A Bacterium that Can Grow by Using Arsenic Instead of Phosphorus [J]. *Science*, 2011: 1163-1166.

Wyrick, J. Steps to Writing Well (Tenth Edition) [M]. Beijing: Peking University Press, 2008.

Yang, S. M.; Kalpakis, K. and Biem, A. Detecting Traffic Events by Coupling Multiple Timeseries with a Bayesian Method [J]. *IEEE Transactions on Intelligent Transportation Systems*, 2014, 15: 1936-1946.

Yarrow, M. J.; Kim Ji Won; Clarkson, W. A. High Power, Long Pulse Duration, Single-frequency Nd: YVO4 Master-oscillator Power-amplifier [J]. *Optics Communications*, 2007, 270: 361-367.

Yu, Z. A. How We Write Papers [J]. *Journal of Analytical Chemistry*, 2006, 6(1): 5.

Zhang, M. Cohesive Features in Exploratory Writing of Undergraduates in Two Chinese Universities [J]. *RELC Journal*, 2000, 31: 61-93.

Pfeiffer William. 科技交流实践教程 [M]. 北京:电子工业出版社, 2006.

余千华,秦傲松. 英语科技论文中的模糊限制语 [J]. 华中科技大学学报(社会科学版), 2001, (11): 121-123.

(http://www.conferencealerts.com; https://index.conferencesites.eu; http://abrcms.org/; http://iafor.org/conference)

附　　录

翻译的启示

　　本人首次翻译英语学术论文还是在大学毕业那年,作为本科毕业设计的一部分而进行的一项任务。现在回想起来,当时翻译的细节已经比较模糊,但非常惊讶于当时的毅力和那股干劲。

　　短短五六年的时间,已经发生了翻天覆地的变化:如今的大学生拥有笔记本电脑已经是非常普遍的事情,而且校园内、公共场合都配备无线网络。此外,3G 和 4G 手机的出现,极大改变了大众的生活,让我们随时随地都可以上网查询几乎所有的信息、上传自己的所见所闻并顷刻间传遍好友圈;而在陌生的地方,手机或 iPad 的 GPS 导航更是令人惬意;更甚者,打的也无须苦苦等待或与他人争抢,只需轻轻一按手机按键,"专车"就到了!

　　然而,就在 2007 年毕业那会儿,没有笔记本电脑,没有无线 Wi-Fi,更没有 3G 手机、iPad mini 等,就连"有道词典"等都闻所未闻。作为一个本科生,还没有进行科学研究,也没读过专业英语论文,当接到翻译的任务时,虽然心中充满恐惧,但却有强烈的翻译欲望。毕竟经过四、六级的连续冲刷和考研英语的洗礼,对于专业英语的翻译,已经潜意识里把它当作一种挑战和享受了!

　　翻译的论文是指导老师打印出来后给我的,而翻译工具是在图书馆能找到的唯一资源——《化学化工英汉词典》,以及一本白皮的不到 100 页的本科专业英语教科书(由学校老师自己汇编的专业书里的章节和注释)《高分子专业英语》。而翻译的内容都是手写在稿纸上的。现在回想起来,翻译的《引发剂残片并入组成自由基聚合法合成二甲基丙烯酸乙二酯的可溶性超支化聚合物纳米粒子》足足花了 5 天的时间,那时我上午 9 点至晚上 6 点都泡在图书馆做翻译工作。刚拿到这篇论文时,第一件事并不是立马翻译,而是先慢慢细读一遍,把认识的专业词汇先标注一下,并粗略估计一下到底能看懂多少。结果,我花了整整半天的时间,却看得懵懵懂懂,甚至有点不知所云。于是我先把它放一放,去把午饭解决了。缓和一下后,再回到漫长的翻译工作当中。

　　由于个人习惯问题,我喜欢按部就班、从头到尾按顺序翻译,但没想到的是,仅题目和摘要,就基本花掉整个下午的时间。这主要由两个方面造成的:一方面,专业英语和四六级、考研英语完全不同,有的句子,特别是摘要部分,读起来特别别扭,好不容易翻译过来后,自己回头读起来要么觉得松散、没连贯性,要么就是太口语化,完全不像专业论文应该有的样子。另一方面,虽然摘要只有短短 100 来个单词,但专业术语高度集中,多种表征方法前所未闻,一些数量单位也不熟悉,而这些术语中,就算到今天网络时代高度发达时,许多专业词汇,特别是新兴词汇,在翻译工具和网络在线翻译里都是一片空白。而当时虽然手头有化学英汉词典和高分子专业词典,也都因为编撰时间太久远,更新不够,以致无从查询!正当手足无措之时,猛然想起图书馆的期刊阅览室。没错,期刊阅览室除了有各种大众媒体期刊,也有不少高校的校报等学术期刊,而令人兴奋的是,阅览室有本专业国内两个最优秀的中文学术期刊《高分子学报》和《高分子通报》,这些期刊上的论文虽然都是中文写就,但都有英文摘要和英文关键词!通过查阅大量过刊和现刊,对比翻译论文的内容,基本上解决了专业词汇的专业化翻译、仪器表征的专业术语翻译,以及数据单位的中文意义等。如此,曲折而繁忙的第一天就过去了。

按照同样的方法,从引言、实验部分、结果与讨论,到最后的结论,都大致一字不差地翻译出来。当然,还有个别实在查询不到或推断出来的原料名称就继续用英文表示了。当把最后的致谢也翻译完后,第一篇论文翻译也就初步完成,这一过程整整花了我五天时间。但定稿还没结束,由于翻译草稿是手写的,所以周末还特意在网吧花了两个多小时做成电子版,然后找地方打印。这样,第一篇英语专业论文的翻译就完成了。

这一经历对我的影响很大,最直接的结果是同寝室的同学和别的宿舍的同学找我帮忙翻译。由于我已经考上了研究生,也没别的事,所以就欣然接受了任务。第二篇翻译花了四天时间,到最后一天就可以翻译一篇。当然,这也是我唯一几次完整翻译英语论文了,在之后的研究生生涯、海外留学研究时,就再没翻译过。只记得读研期间,帮师弟师妹们口头翻译过少数句子和段落。

也许正是本科毕业那段翻译的经历,让我在研究生学习伊始就习惯看英文文献,我不但没有恐惧心理,反而觉得理所当然。如今,六年过去了,我已经发表了 20 余篇论文,全是英语论文并发表在国际期刊上,比较遗憾的是,我没有写过一篇中文论文。

王　锦
中国科学院苏州纳米技术与
纳米仿生研究所